shared feeling that "the artist creates not what others think is beautiful but whatever is necessary for him," the book contains articles by Marc, Schönberg, Delaunay, and others of the *Blaue Reiter* group, an original play by Kandinsky, along with his famous essay "On the Problem of Form," and it is richly illustrated with paintings by the editors and with a large selection of works they most admired: medieval, primitive, Japanese, folk, children's art, as well as such modern masters as Cézanne, Rousseau, and Van Gogh.

Published in 1965 in a German documentary edition by Klaus Lankheit, the present translation includes Lankheit's extensive critical introduction, bibliographical and biographical notes—edited and supplemented by Bernard Karpel—and essential commentary on the collaborators and editors.

# The *Blaue Reiter* Almanac

# THE DOCUMENTS OF 20TH-CENTURY ART

ROBERT MOTHERWELL,
GENERAL EDITOR

BERNARD KARPEL,
DOCUMENTARY EDITOR

ARTHUR A. COHEN,
MANAGING EDITOR

# THE *BLAUE REITER* ALMANAC

## EDITED BY
## WASSILY KANDINSKY
## AND FRANZ MARC

NEW DOCUMENTARY EDITION
EDITED AND WITH AN INTRODUCTION BY
KLAUS LANKHEIT

NEW YORK    THE VIKING PRESS

*Der Blaue Reiter* © R. Piper & Co. Verlag, München 1965
English-Language translation © 1974 by
Thames and Hudson Ltd, London

Published in 1974 in a hardbound and paperback edition by
The Viking Press, Inc., 625 Madison Avenue, New York, N.Y. 10022
Published simultaneously in Canada by
The Macmillan Company of Canada Limited

SBN 670-17355-x (hardbound)
    670-01931-3 (paperback)

Library of Congress catalog card number: 75-170678

Printed in Germany by Herder Druck GmbH Freiburg
Bound in Holland

Translated by Henning Falkenstein,
with the assistance of Manug Terzian and Gertrude Hinderlie

# Contents

Appendix

# Preface to the English-Language Edition

The art of Wassily Kandinsky and Franz Marc cannot be appreciated without taking note of the considerable number of their works in museums and private collections in the United States. Personalities such as Arthur J. Eddy and Hilla Rebay were early collectors of important paintings by the two artists.

The recent demand for an English translation of the theoretical principles of these painters is only logical. It proves that *Der Blaue Reiter* has become a classic document of twentieth-century art and that its significance has not decreased.

I have taken the opportunity of the publication of an edition in English to correct and supplement the German edition of 1965; works that were published subsequently have been incorporated.

*Karlsruhe, Winter 1970-71*                                    K. L.

# Preface to the 1965 German Edition

More than fifty years ago, in 1914, the second and until now the last edition of the anthology *Der Blaue Reiter* (The Blue Rider) appeared; the first edition was published in 1912. The second edition coincided with the tenth anniversary of Piper Verlag, in Munich, the publisher of the volume. Printed like the first in a small edition, this second edition has long been out of print.

Since then the volume has gained the reputation of being the most important programmatic work of twentieth-century art. This is unanimously recognized by scholars throughout the world: "The almanac remains unique among European writings on art; no other country has produced a comparable work capturing the excitement and tension of the years before World War I" (Will Grohmann). "The *Blaue Reiter* was a movement of truly historic significance and fruitful power . . . its artistic insights, which were formulated at that time, are still as valuable today as they were half a century ago" (Marcel Brion).

A new edition, therefore, does not require any justification; it is one of the most urgently needed works of literature on art theory and culture.

Although it was easy to arrive at the basic decision to publish a new edition, questions concerning the format of the book were more difficult to resolve. Manuscript notes of Kandinsky and Marc show that they were not fully satisfied with the contents and make-up of the almanac; they very probably would have made changes for their third edition. Another editor of a third edition, however, cannot possibly do this. Today we value the *Blaue Reiter* as a documentary source—and it is essential to preserve this quality. A facsimile edition, however, was out of the question. Due to the quality of the paper, the illustrations of the first two editions left so much to be desired that a new edition could hardly have rectified it. Second, even the most exact imitation could not have re-created the atmosphere that, in a strange way, breathes through the original. Also, the cost of production, especially of the binding, would have been so high that only a very small number could have afforded it. So the book would not have reached the kind of reader for whom it was primarily intended.

Publisher and editor, therefore, decided to publish a "study edition." Size, typography, and paper were modernized, but each word and each illustration of the original were preserved; nothing was left out and nothing was added. The sequence of texts and illustrations was retained. Only printing errors were corrected and punctuation adjusted to modern rules. This present edition is based on the first one; prefaces to the second edition are made part of the appendix. Since the make-up caused changes, numbers in brackets are included to indicate the original page numbers, so that the reader can envision the original layout.

Additional documentation and a critical commentary, however, could not be dispensed with: after half a century, many events of the art world of the period and many names familiar to contemporaries are now forgotten. Many factual errors, unavoidable considering the state of knowledge at that time, had to be corrected. Special care was given to exact information about the illustrations—which has always been incomplete until now—and to their present location; whenever possible, new plates were procured. Finally, the modern reader is entitled to know the essentials about editors and collaborators, about the artistic and spiritual background, about the circumstances surrounding the origin of the publication, and about its subsequent effect.

"There are no documents about the days in which the famous anthology was created." This opinion of August Macke's biographer is still prevalent but nonetheless incorrect. The sources for the decisive months, weeks, and days of the genesis of the almanac are abundant. But these unique resources, the entire correspondence of Wassily Kandinsky and Franz Marc between the years 1911 and 1914, have to this day been neither published nor used. To complete the picture we must also consider the estate of Reinhard Piper, which includes numerous additional testimonies and discoveries from the Old World and the New. Using this authentic material and taking advantage of new discoveries, I hope to encourage future research.

The present edition has been made possible only through great understanding and support for the publisher and editor. I am especially thankful to all those people who without restriction gave us permission to reprint texts protected by copyrights and who also provided me with much additional information: Mme. Nina Kandinsky, Mrs. Gertrud Schönberg, Mrs. Olga von Hartmann, Rudolf Probst, and last, but not least, David Burliuk, the still actively working fellow rebel of the *Blaue Reiter*.* I also am indebted to collectors, directors of museums and galleries, and the church administration, who not only permitted the reproduction and reprinting of works of art in their custody but also supported the publication with valuable information about them: the Catholic Parish Office in Murnau, Upper Bavaria, the administrations of the Historisches Museum of Bern, the Völkerkundemuseum of Munich, the Heimatmuseum of Oberammergau. The author wishes to thank all those persons who assisted him with individual advice, biographical or bibliographical hints, or in procuring photographs; particular appreciation goes to the late Frau Maria Marc, Mme. Sonia Delaunay, Rose Hempel, Maurice Besset, the late Bernhard Koehler, Jr., the late Wilhelm Worringer, Ludwig Grote, Daniel-Henry Kahnweiler, Wolfgang Macke, Aimé Maeght. I am indebted to the Library of Congress, Washington, D.C., the Staatsbibliothek Stiftung Preussischer Kulturbesitz, Marburg, and the Kubin Archive, Hamburg, for their permission to use or mimeograph texts. Herr Klaus Piper has opened the archives of his publishing house as well as his own private collection of the documents of Reinhard Piper.

My thanks are also due to the publishing house of DuMont Schauberg in Cologne, which generously permitted the advance printing of some important passages from the correspondence of Kandinsky and Marc. The undersigned is preparing a complete edition of this correspondence, hoping to present the first comprehensive representation of the *Blaue Reiter* movement based on the sources. All questions that would be beyond the scope of an already far-reaching commentary to this edition will be included in the edition of the correspondence. Important scholarly sources are given in footnotes and short biographies.

*Karlsruhe/Heidelberg, Summer 1965*                                          K. L.

---

*David Burliuk died in 1967, two years after this preface was written. (Trans.)

# A History of the Almanac
*by Klaus Lankheit*

## Early History

In 1911 the firm of Albert Langen in Munich published a small volume by Victor Aubertin with the alarming title *Die Kunst stirbt* (Art Is Dying). The author presented his theses sharply and aggressively. "Art," he wrote, "is dying of the masses and of materialism. It dies because the land it needs is all built up, the land of naiveté and of illusions. . . . On each national holiday a joint toast to art and science is proposed; perhaps they mean one and the same to the idiot. But they are deadly enemies: where one of them exists, the other flees." Immediate and rousing as these words may still sound today, the following statement is surprising: "We must confess that we no longer have an artistic idea . . . for the first time we have entered a period without direction, without an artistic style, without a young revolutionary generation."

The years before World War I may be described in many ways, but to say that they lacked a young generation in revolt against the established art seems absurd to us today. Everywhere there was unrest, everywhere groups, both large and small, were formed in order to give their views greater force in the fight against traditional trends. Munich (where since 1896 the two satirical magazines *Jugend* [Youth] and *Simplizissimus* had been giving free rein to their mockery) was already —during the same period, in 1892—the locale of the first of the "secessions" of the German-speaking world, which had been founded to challenge the academy. But by 1911 the revolutionaries of that earlier time were established professors. New groupings like the Scholle (Soil) suffered a similar fate of stagnation. In 1909 Wilhelm Worringer, a young Ph.D. in art history, criticized mercilessly: "The Secession has become fat. In spring and summer there is the same atmosphere of saturation, the same crystal-palace *ambiance*, the same marketable works. In order to conceal bourgeois portliness, one sends for some sensations from Paris that would never pass the jury through regular channels, i.e., if they had come from a Schwabing studio. . . ."*[1]

Worringer obviously knew about new forces establishing themselves in the Bavarian capital. This was the dawn of the "visionary advent," which Hugo Ball was to experience shortly thereafter. At the beginning of the same year, 1909, the Neue Künstlervereinigung München (New Artists' Association of Munich, or NKVM) had been founded in Schwabing. Later, in the *Blaue Reiter*, Franz Marc wrote, looking back on it: "In Munich the first and only serious representatives of the new ideas were two Russians who had lived there for many years and who had worked quietly until some Germans joined them. Along with the founding of the association, those beautiful, strange exhibitions arose that drove critics to despair." The two Russians referred to by Marc were Wassily Kandinsky and Alexei von Jawlensky. From personal experience Kandinsky described the aims and genesis of this association:

> The membership list included international names—Germany, France, Austria, Russia, Italy. . . . Apart from this union of various countries to serve one purpose, which we all thought to be supreme, there was one other aspect of the alliance that was new at that time: the members chosen, in addition to painters and sculptors, were also musicians, poets, dancers, and art theoreticians. In other words, we wanted to unite the individual trends that up to this time had been separated both externally and internally. . . . Without a *deus ex machina*, however, all these enthusiastic plans would have been futile; for we needed an art dealer for our exhibitions and we did not find one. At this critical moment we learned that Geheimrat [Privy Councilor] Hugo von Tschudi had been appointed director of all Bavarian museums. Of course, asking this important man for help was rather bold (some called it inconsiderate, others impudent). However, he was not only an important but also a great man. . . . At that time Heinrich Thannhauser was believed to have the best exhibition rooms in all Munich, and Tschudi extorted these rooms from him. . . . The press raged against the exhibition, the public railed, threatened, spat at the pictures. . . . We exhibitors could not understand this indignation. . . . We were only amazed that in Munich, the "city of the arts," nobody except Tschudi had given

---

*Schwabing is the artists' quarter of Munich. (Trans.)
[1]*Kunst und Künstler*, VIII (1909), 369ff.

12

us a word of sympathy. And one day this word came. Thannhauser **History** showed us a letter from a Munich painter who was then unknown to us. He congratulated us on our exhibition and proclaimed his enthusiasm in most expressive terms. This painter was a "genuine Bavarian," Franz Marc.[2]

In his polemic, which was animated with buoyancy and enthusiastically worded, the thirty-year-old Marc praised the "fully spiritualized and dematerialized inwardness of feeling" of those works. Both his phrases and his indignant newspaper pieces express the revolutionary spirit kindled by the exhibitions of the New Artists' Association.[3] How much the masters of the *Blaue Reiter* owed to developments in the years 1909 and 1910 is too often forgotten. It was in Munich of all towns that the breakthrough of modern art on the eve of World War I occurred in two waves that are characterized by the Neue Künstlervereinigung and the *Blaue Reiter*. The second could not have taken place without the first. This should especially be kept in mind as the dissolution of the association was historically inevitable. At first, however, the members of the association unanimously welcomed his help in publicizing them; they invited Marc to join them and elected him a member of the board.

But soon afterward, as early as the spring of 1911, slight tensions within the group became noticeable. They were based on insurmountable human and artistic conflicts. After temporary compromises the tensions increased, because the jury held differing opinions. The speakers of one faction were Adolf Erbslöh, the chairman, and Alexander Kanoldt; the other faction was represented primarily by Kandinsky and Marc. On August 10, Marc described the situation in sharp words to his friend August Macke: "Together with Kandinsky I foresee clearly that the next jury meeting (in late fall) will bring about a horrible argument that then or next time will result in a split or in the resignation of one of the two factions; the question is, which of the two will *survive.* . . ."[4]

The predicted separation occurred a few months later. On December 2, the two friends left the association; other members, such as Gabriele Münter, Alfred Kubin, Thomas von Hartmann, and Henri Le Fau-

[2]*Franz Marc im Urteil seiner Zeit:* Introduction and Commentary by Klaus Lankheit (Cologne, 1960), pp. 45f.
[3]*Zur Ausstellung der Neuen Künstlervereinigung bei Thannhauser.* Special edition (1910).
[4]August Macke and Franz Marc, *Briefwechsel* (Cologne, 1964), p. 65.

connier declared their solidarity with them. The final cause was the fact that the jury of the association had rejected Kandinsky's *Composition No. 5* for its winter exhibition. One day after the split Marc wrote to his brother:

> The die is cast. Kandinsky and I . . . have left the association. . . .
> Now it is the two of us who must continue to fight! The Editors of the *Blaue Reiter* will now be the starting point for new exhibitions. I think this is quite good. We will try to become the center of the modern movement. The association may assume there the role of the new Scholle.[5]

Many years later Kandinsky described the counteraction much more extensively:

> Since the two of us sensed the "crash" sometime beforehand, we had prepared a different exhibition. Thannhauser gave us his rooms — right beside the NKVM's exhibition rooms. In the catalogue of our exhibition we stated that this exhibition did not intend to propagate a definite trend but was rather based on the principle of a variety of artistic expressions. The "front" comprised a "left" wing (the newly created "abstract" side) and a "right" wing (purely "realistic"). Such a synthesist basis was new at the time, and has remained new ever since. As the organizers of the exhibition, we named the Editors of the *Blaue Reiter*, since we were already working on this book.[6]

The "First Exhibition of the Editors of the *Blaue Reiter*" was put together, in a frantic hurry, and it opened on December 18, 1911. Later it traveled all over Germany, and as late as 1914, Herwarth Walden of the Berlin Sturm Gallery took it abroad, after some works had been changed several times. While the exhibition was still in Munich, Marc and Kandinsky began planning another one. This "Second Exhibition of the Editors of the *Blaue Reiter*" took place in March of 1912 at the Goltz gallery; it was restricted to water colors, drawings, and printed graphics. There is no history of twentieth-century German art that does not give ample room to a description of those two exhibitions, and it is therefore unnecessary to re-evaluate them

---

[5]Klaus Lankheit, "Zur Geschichte des Blauen Reiters," *Der Cicerone*, III (1949), 110ff.
[6]*Franz Marc im Urteil seiner Zeit*, p. 48.

here. In the present context one must rather point to the facts given by Marc and Kandinsky themselves. The exhibitions were organized by the "Editors" of a proposed book that was to bear the title *Der Blaue Reiter*. Accordingly, *Der Blaue Reiter* was originally nothing but the title of a book that was transferred to the exhibitions.

In his account in 1935, Kandinsky continues: "There never really was a *Blaue Reiter* society, not even a group, as is often incorrectly stated. Marc and I took what we thought was good; and we selected freely without considering certain opinions or wishes." And with ironic scorn, Kandinsky says: "So we decided to run our *Blaue Reiter* 'dictatorially.' The 'dictators' of course were Franz Marc and myself."

In the history of art, however, exhibitions and book belong together. They were based on the same convictions and carried out according to the same principles. Chronologically, as well, the two undertakings ran parallel. Thus it is not without good reason that the name *Blaue Reiter* is still applied to both book and exhibitions. As most of the artists whose paintings were being exhibited at the Thannhauser and Goltz galleries were also collaborating on the almanac, the name could now be applied to a larger circle of like-minded artists. But the *Blaue Reiter*—in its two forms—was ultimately the very personal achievement of two congenial individuals.

## The Plan of the Almanac and the Work of the Editors

As early as June 1911—that is, *before* the situation had become critical and the friends had decided to leave the New Artists' Association— Kandinsky conceived the idea of an almanac. On the nineteenth of that month he presented Marc with the plan:

> Well, I have a new idea. Piper must be the publisher and the two of us the editors. A kind of almanac (yearbook) with reproductions and articles . . . and a *chronicle*!! that is, reports on exhibitions reviewed by artists, and artists alone. In the book the entire year must be reflected; and a link to the past as well as a ray to the future must give this mirror its full life. The authors will probably not be remunerated. Maybe they will have to pay for their own plates, etc. We will put an Egyptian work beside a small Zeh [the

15

last name of two talented children], a Chinese work beside a Rousseau, a folk print beside a Picasso, and the like! Eventually we will attract poets and musicians. The book could be called "The Chain" or some other title. . . . Don't talk about it. Or only if it could be directly useful to us. In cases like this "discretion" is most important.

This letter of June 19, 1911, is the birth certificate of the *Blaue Reiter*. Kandinsky, who was much older and more experienced than Marc, must be credited with the initial idea, which already included important basic ideas for the publication: the two painter friends as editors, the selection of authors from all the arts, the inclusion of the latest foreign works as well as Egyptian and East Asian art, folk art, children's art, and amateur paintings. The underlying principle of comparing and contrasting works from different areas and periods must be particularly noted. A synthesis of the arts would be stimulated by including literature and music. Finally Kandinsky makes practical proposals for the realization of the plan: he specifies a publisher and is convinced that the considerable cost of production could be markedly reduced by saving on fees and the cost of plates.

Franz Marc cheered this plan enthusiastically. He possessed all the qualities of a born co-editor. As his brother recalled, he had talked about a magazine of his own as early as 1910. His support of the Neue Künstlervereinigung had established his name as an excellent writer on art problems of the day. Kandinsky could not therefore dispense with the "delicate, understanding, and gifted spiritual co-operation and assistance" of his friend; and Marc joyfully contributed his talent for organization and his author's pen.

The following weeks and months were filled with the feverish activity of the two "editors." A letter of Kandinsky's dated September 1, 1911, colorfully depicts the bubbling abundance of their ideas, their enthusiastic energy, and the very impressive scope of their aims.

. . . I for my part wrote to Hartmann, told him about our union, and bestowed on him the title of "Authorized Representative for Russia." I asked him expressly to feel with all his soul what that meant. I am going to write also to Le Fauconnier. . . . From Hartmann I ordered an article on Armenian music and a music letter from Russia. . . . I received a copy of the manifesto of the Italian Futurists, which gives us some material on the Italian musical movement. Schönberg *must* write on German music. Le Fauconnier *must*

get a Frenchman. Music and painting are already properly covered **History**
There should also be a few scores. Schönberg, e.g., has lieder. We
might also ask Pechstein to write a Berlin letter: nothing too definite,
just to try him out. Miss Worringer on the Gereon Club and its
aims. Just wait! We will get a real pulse flowing in our dear book.
We also have permission to include parts of Tschudi's "Gallery
Director." We must show that something is happening *everywhere*.
We will include some reports on the Russian religious movement in
which *all* classes participate. For this I have engaged my former
colleague Professor Bulgakov (Moscow, political economist, one
of the greatest experts on religious life). Theosophy must be
mentioned briefly and powerfully (statistically, if possible). . . .

The ties were to reach from Paris to Moscow, and from Berlin to
Milan; music was to be represented on as international a basis as
painting; art was to be face to face with religious movements. Kandinsky's friendship with the Russian composer Thomas von Hartmann
and his acquaintance with Arnold Schönberg of Vienna guaranteed a
good representation of music. Henri Le Fauconnier, a member of
Munich's Neue Künstlervereinigung and also vice president of the
Indépendants, was the right man to report from France. Max Pechstein was considered the leader of the Neue Sezession (New Secession)
in Berlin, which, since 1911, had gathered all revolutionary elements—
especially those from the former Brücke. Hugo von Tschudi, director
of all Bavarian State Galleries, who was already hopelessly ill, had
previously supported the exhibitions of the NKVM; he had again
promised support to the two editors, who visited him frequently.
Tschudi had written a preface to the catalogue for an exhibition of the
Hungarian collector Marczell von Nemes in the Alte Pinakothek,
where eight paintings by El Greco had given the German public their
first impression of the master. In it Tschudi proposed a new type of
museum director—and Kandinsky wanted to print a paragraph from
this essay.[7] Emmy Worringer, sister of the art historian, had founded
the Gereon Club in Cologne to promote modern art; she was to show
the "First Exhibition of the Editors of the *Blaue Reiter*" in Cologne in

[7]On Tschudi cf. *Gesammelte Schriften zur neueren Kunst von Hugo von Tschudi*,
E. Schwedeler-Meyer, ed. (Munich, 1912). (In this volume cf. especially E.
Schwedeler-Meyer, "Biographische Skizze," pp. 9ff., and "Vorwort zum Katalog
der aus der Sammlung Marczell von Nemes—Budapest in der Kgl. Alten Pinako-
thek zu München 1911 ausgestellten Gemälde," pp. 226ff.); K. Martin, *Die
Tschudi-Spende: Hugo von Tschudi zum Gedächtnis* (Munich, 1962).

January 1912. The exhibition was ridiculed there, on its first showing outside of Munich.

During the fall the editorial work goes on in Murnau, Sindelsdorf, and also in Munich. Schönberg pays them a visit. David Burliuk, who had written an article in the 1911 catalogue of the NKVM, is asked for a contribution; Le Fauconnier has to be reminded; from Paris Kahnweiler sends photographs of Picassos. Matisse is asked for permission to reproduce his works and to write an essay: they could publish whatever they wanted, he replied, but writing was impossible because "one must be a writer to do something like that." In early September, Marc finishes his article "Two Pictures"; four weeks later "Spiritual Treasures" is ready. At the same time Kandinsky works on his introduction "On Stage Composition," and he is already preparing the text of the press release. The first disappointments occur: changes must be made. Sometimes amid the enthusiasm slight doubts creep in. After a conference with the publisher, Kandinsky writes on September 21: "I feel a little funny. Just like . . . well! like before an attractive, tremendously interesting mountain climb, but one where you have to crawl through crevasses and ride on ridges."

The term "Der Blaue Reiter" must have been coined during this period. This strange title caused much speculation as to its origin. But Kandinsky himself commented on the name in 1930, when Paul Westheim asked for an explanation: "We invented the name 'Der Blaue Reiter' while sitting at a coffee table in the garden in Sindelsdorf; we both loved blue, Marc liked horses, I riders. So the name came by itself. And after that Mrs. Maria Marc's fabulous coffee tasted even better."[8]

Even after two decades the artist could still clearly see that afternoon with the Marcs in his mind's eye down to the last detail. The very

[8]W. Kandinsky, " 'Der Blaue Reiter' (Rückblick)," *Das Kunstblatt*, XIV (1930), 59n. In the secondary sources an explanation for the name has been handed down in an attempt to link the almanac with an earlier painting of Kandinsky's; according to this explanation the artist carried over the title of the painting to the book eight years later. There actually is a painting from the year 1903 showing a rider in a landscape (Switzerland, private collection). But the content of this early painting leads to a different, i.e., personal sphere. It is an almost square canvas. The influence of Jugendstil is obvious. As with Ferdinand Hodler's paintings there is a sloping sunny meadow: the dimension of depth is forfeited in favor of a plane background. In their parallel structure the trunks of the autumn birches accentuate the rider in a blue coat, who is chasing his own shadow while riding a white horse. In this chalky white coloring and the nervous palette-knife technique lies something threatening that points to an inner unrest, even uncertainty. This

directness of the report, which is as unsentimental as possible, indi-
cates that the "historic" incident really occurred without any ceremony
and in great friendship. Kandinsky's report is backed up by documents.
In the correspondence of the two friends we find the term "Der Blaue
Reiter" for the first time on September 21, 1911.

The provisional table of contents, as we shall see, does not contain
the name of August Macke. He first learned about the project in a
letter from Marc dated September 8.[9] For some time, however, he
seems to have been expressing ideas that would make him a valuable
collaborator. He must have had previous ideas of his own, especially
on "comparative history of art," and furthermore he was the only one
who had practical experience with the theater, dating from his work
at the Düsseldorf Theater. Therefore it is not surprising that the
invitation to collaborate fell on fertile soil. On the twenty-fifth of the
month he wrote from Bonn to Gabriele Münter:

> I compiled a few important things which will be hard to abandon.
> "Justification of Peasant Art," or "Character in Pottery Ornaments,"
> "Artistic Trends in African Secret Societies," "Masks and Puppet
> Plays among the Greeks, Japanese, Siamese," "Mystery Plays
> among Heathens and Early Christians," "Living and Dead Orna-
> ment," "Bare Facts in Art," etc. All these things make my head
> bubble in confusion. Whenever I pick out something intelligent,
> I like to write it down. By the way, I will soon go to Munich, Sindels-
> dorf, and Murnau.[10]

In early October, Macke did arrive in Sindelsdorf and was intoxicated by the same enthusiasm that had seized the friends. He asked his

painting predates the period of the almanac's publication, which was marked by
prophetic self-confidence and high spirituality, although it is true that the painting
is now called *Der Blaue Reiter*, and that it bears the same title in the master's
handwritten catalogue of his works. Through the generosity of Nina Kandinsky,
I had the opportunity to study this catalogue, which was part of his estate, in
Paris; it revealed that Kandinsky's original pencil notes on this very page had
grown so pale that he later rewrote them in ink. Underneath the ink the original
title is still visible; the painting was called just *Der Reiter*. Therefore the development must have taken the opposite direction from what scholars have thought:
the 1903 painting did not give the almanac its title; during the revision of his
catalogue of works, Kandinsky was reminded of the book and the movement it
caused, and following a whim, he renamed the picture. So we have removed the
final cause to doubt the artist's report to Paul Westheim in 1930.

[9]Macke and Marc, *Briefwechsel*, pp. 72f.
[10]Quoted from *August Macke*, exhibition catalogue (Munich, 1962), p. 63.

wife to join them and raved: "All the days are like holidays. . . ."
Soon Marc could report to Kandinsky: "August is working on his
article." During their common talks, out of the many themes bubbling
in Macke's head his poetical essay "Masks" had crystallized, and
pictures from the Völkerkundemuseum in Munich were selected to
make it more attractive.

The picture section was expanded considerably when Marc became
acquainted with the abundant creations of the Brücke group, which,
as Kandinsky admitted in 1930, "was completely unknown in Munich."
Marc spent New Year's Day with his parents-in-law in Berlin and
visited Pechstein, Kirchner, Heckel, Müller, and Nolde. Inspired by
this atmosphere, he reported almost daily to Munich; when Kandinsky
expressed some doubt or objection about this art, Marc became more
and more eloquent. His penetrating judgment of these masters in his
letters from the capital is the most beautiful and also among the
earliest testimonies on the Brücke. Marc was immediately impressed
by the "truly artistic air," by the "very powerful things throughout,"
and by "a tremendous untapped wealth that is as close to us as the
ideas of our silent admirers in the land"; he found a "gigantic stock of
material" for the planned second exhibition of the *Blaue Reiter*, "in
which they would like to participate without any pretensions."

Today the graphics of the Brücke are considered "one of the most
important of Germany's contributions to modern art" (Peter Halm).
But Kandinsky's reservations did not vanish completely; he wrote to
Marc on February 2:

> . . . Those things must be *exhibited*. But I think it is incorrect to
> immortalize them in the *document* of our modern art (and, this is
> what our book ought to be) or as a more or less decisive, leading
> factor. At any rate, I am against *large* reproductions. . . . The small
> reproduction means: this *too* is being done. The large one: *this*
> is being done. . . .

Occasional differences of opinion and divergent artistic judgments
could not be avoided between two such powerful personalities, but
disagreements could never shake their mutual trust, even if "the
strange hypersensitiveness of Münter"[11] should at one time put it to a
severe test. The great enthusiasm and the great hopes, as well as the

---

[11]Elisabeth Erdmann-Macke, *Erinnerung an August Macke* (Stuttgart, 1962), p. 189.
   Cf. also Macke and Marc, *Briefwechsel*, pp. 110ff.

unexpected setbacks and forced compromises in the editors' work, are best reflected by the changes of the program during the decisive months. Notes, announcements, press releases, and prospectuses from the various phases give an insight into the difficulties of the enterprise.

In a letter to Reinhard Piper dated September 10, 1911, Marc included a "provisional table of contents of the first number" (Documents, p. 243). It is clearly structured. Each preface is followed by four main sections of the text: painting has six individual contributions, music eight, the stage three, and the chronicle of daily events two. The middle section is particularly strong: after an introduction by Kandinsky and Schönberg's contribution, five authors were to write on Russian movements and one on French music. Of the essays on painting, only those by Marc and Burliuk and—with a different title —the one by Kandinsky were finished in time. Similarly, only three essays on music were handed in, with two authors changing the subject. Of the three essays on the stage, only the one by Kandinsky appeared, and the "chronicle" itself was completely canceled. Striking differences between plan and final choice also occurred with the reproductions. The *Images d'Epinal,** examples of the French folk prints that Girieud wanted to make available, did not arrive in time. No mention is made later on of reproductions of Max Oppenheimer's works. Jawlensky and Werefkin were rejected by the bitterly disappointed Marc and Kandinsky, after they had let them down during the decisive meeting of the Neue Künstlervereinigung and taken sides with Erbslöh and Kanoldt.

The next table of contents is preserved in two slightly different versions. One is the press release for the almanac in Kandinsky's treatise, *Concerning the Spiritual in Art*, and the other is Kandinsky's manuscript in his own hand, which later was also used as a supplement to the exhibition catalogue (Documents, p. 246). It could have been written in early October 1911, but it seems to have been altered again because the last part was obviously tacked on later. In order to give an idea of the program, painters, musicians, poets, and sculptors(!) are mentioned as collaborators. The nine contributions "from the first volume" present very considerable changes compared to the provisional table of contents. August Macke and Roger Allard have taken the places of the rejected authors. Marc announced another essay.

---

*I.e., *Pictures from Epinal*, produced in the nineteenth century by citizens of Epinal, France, showing mostly patriotic battle scenes and popular heroes. (Trans.)

In the printed version, which was completed with the first edition of *Concerning the Spiritual in Art* at Christmas 1911, Pechstein is— still or again—represented, as well as the Russian musicologist N. Brüssov. Kandinsky's topic is still "Construction"; Schönberg's article does not yet have its final title, but it had been changed to "The Question of Style." Most striking is the great expansion of the picture section, which is to contain about a hundred items. Besides folk art, we now see primitive art, classical art, and children's art. The present, too, is much better represented: Van Gogh (d. 1890), Gauguin (d. 1903), and Cézanne (d. 1906) are added to their own "twentieth century" without hesitation; Matisse is included among the younger painters. The later handwritten version of the table of contents no longer contains the names Jawlensky and Werefkin, so it could hardly have been written before December 2; but Kandinsky's treatise was already in type—therefore those names remained in the press release. Finally, musical examples are mentioned for the first time: lieder by two disciples of Schönberg, Alban Berg and Anton von Webern.

The most important announcement of the book was the four-page subscription prospectus. It dated back to a conference between Kandinsky and the publisher on November 24, 1911, but it did not appear until February of the next year. Marc wrote the text in January during his stay in Berlin (Documents, p. 252). In the list of contents the Pechstein article is definitely deleted, but newly included are contributions by E. von Busse on Delaunay, and by L. Sabaneiev on Scriabin. Schönberg specified his topic and promised music of his own. Kandinsky's main article already had its final title. In the reproduction section, which is enlarged by twenty items, the name of Henri Rousseau, the great discovery of the pre-Christmas period, is added.[12]

A small announcement adorned with Rousseau's painting was probably printed in early March and then appended to the catalogue of the *Blaue Reiter* exhibition at the Berlin Sturm Gallery, which opened on the twelfth of that month (Documents, p. 256). The number of pictures has increased again by twenty to "about a hundred and forty." The number of reproductions now corresponds to those of the list in the book. Both Marc and Kandinsky had written another article, and the Russian Kulbin had submitted a new one on "Free Music."

[12]This subscription prospectus is referred to in a full-page announcement in the *Börsenblatt für den Deutschen Buchhandel*, March 2, 1912.

At a meeting between Kandinsky and the publisher on April 5, the Brüssov article was deleted at the last moment in order to keep the size within limits.

Ultimately there were fourteen larger articles interspersed with shorter notes and quotes. In the middle of May the long-expected volume appeared and was greeted by the editors with joy, but also with some signs of annoyance. Both received their complimentary copies on the same day and both immediately wrote letters to each other that reflect their respective temperaments:

> Dear Marc, What do you say about Piper? He did not improve a thing!! Tschudi is printed on the back page, one mirror painting cut off, etc., etc. These are only trifles, but it is outrageous and a shame all the same (especially about Tschudi). Are you perhaps going to write a stiff letter of complaint?

> Dear Kandinsky, Today the expected shipment of ten complimentary copies of the *Blaue Reiter* arrived from Augsburg, but without the last-minute changes (dedication to Tschudi, Egyptian print, etc.). Did Piper not accept the alterations or did he hold them back on his own initiative? Now that I am confronted with the *fait accompli*, I am getting used to it, at least I have to. The impression of the book is after all fabulous. I was so happy to see it before me finally finished. I am sure of one thing: many silent admirers in the land and many young forces will thank us secretly, they will be enchanted with the book and will judge the world by it. . . .

## Reinhard Piper and the Printing of the Book
## The Patron Bernhard Koehler

"Piper must be the publisher," Kandinsky had already written in the letter that first developed the plan for the almanac. It was obvious that no other publisher could be considered. In fact, whoever writes the history of the *Blaue Reiter* is obliged also to devote attention to the personality and work of Reinhard Piper. He will not only discover why the volume was published by this publisher but also receive a condensed introduction into the cultural and especially the artistic

23

Munich publishers, as seen by a contemporary caricaturist: Reinhard Piper (second from right), Georg Müller (far left), and R. Oldenbourg (fourth from left).

milieu out of which the *Blaue Reiter* grew. For, as Ernst Penzoldt put it, the house of Piper "essentially participated in the bloodless revolution in art that led from the nineteenth-century Wilhelmine epoch to the modern era." While they were editing, the two impatient artist friends were sometimes, to be sure, annoyed by the economic restrictions imposed by the publisher and in their correspondence gave vent to their feelings in drastic words (what author would not do this occasionally!). But on the other hand, we find Kandinsky writing to Marc on September 18, 1911: "Piper is the best company after all, and a Munich one to boot." Reinhard Piper's place in history is to have made it possible for the *Blaue Reiter* to be realized. It was primarily this volume that gave the house the odium of being revolutionary. In a cartoon portraying Munich publishers, Piper was shown with the almanac in hand as his most significant characteristic (see illustration). He mentions this episode in his memoirs, which survey his work as well as his personal relationship with Marc and Kandinsky. We sense the fateful meeting of the two: "It was almost natural that Marc offered this book to me."[13]

[13]Reinhard Piper, *Vormittag: Erinnerungen eines Verlegers* (Munich, 1947; new ed., 1964). Klaus Piper, ed., *Nach fünfzig Jahren. Almanach* (Munich, 1954). The quotation by Ernst Penzoldt is from this book, p. 25.

Reinhard Piper has described vividly how he first met Marc. It was at the beginning of 1909:

> While I was working on the book *Das Tier in der Kunst* [The Animal in Art], the still unknown Franz Marc had his first exhibition at Brakl's in Goethestrasse. Most of his pictures showed animals, done predominantly in the impressionist style. The average price was two to three hundred marks. I bought the colored lithograph with two horses. The sales were so slow that even the smallest one was noticed by the artist. He therefore visited me at the office, slim and dark-haired. He came just in time for me to be able to include, on the next-to-last page of my book, one of his sculptures, a group of horses in bronze.

Out of this first meeting a rather active relationship developed. "He was very interested in my publishing. For the cover of my *Das Tier in der Kunst* he made a powerful black-and-white sketch of Delacroix's water color of the horse frightened by lightning; he did the same thing with Cézanne's *Women in Front of a Tent* for Meier-Graefe's short monograph on that artist."

Marc's own activity in the Piper publishing company was sparked by the contemporary art scene and as such served as a starting point for the *Blaue Reiter*:

> At that time Eugen Diederichs published an essay by the Worpswede painter Carl Vinnen protesting the purchase of a Van Gogh by the Bremen Kunsthalle; the essay attacked the supposed overrating of French painting and the activities of progressive museum directors in general. Vinnen had rounded up a number of painters who felt affected by this overrating. . . . Hugo von Tschudi, the courageous reformer at the National Gallery, had been ousted from Berlin and had just started his work at the Munich Pinakothek. The attack was directed against him and also against Meier-Graefe's activity. A rebuttal was unavoidable. Marc and others were busy gathering contributions from important persons. This resulted in the essay "German and French Art." Its significance was much greater than its cause, and it has remained an important document of contemporary history.

In addition to prominent museum directors and collectors, almost all ranking German artists are represented; besides Worringer there are the two creators of the *Blaue Reiter*. Kandinsky, too, had been

25

acquainted with Reinhard Piper for some time. For Piper's onetime "Graphics Sales Department" he had supplied "colored fairy-tale woodcuts." However, the publisher remained wholly divorced from the artist's later magnificent development; a few sentences in his memoirs, written after World War II, reveal that he had still not found contact with the values of this revolutionary art.

But now—in the fall of 1911—on Marc's recommendation, Piper endorsed Kandinsky's manuscript, *Concerning the Spiritual in Art*, which, according to its author, "had been vainly seeking a publisher for almost two full years." Piper printed it so quickly that it appeared in time for the "First Exhibition of the Editors of the *Blaue Reiter*." He backed the work of the two editors as strongly as he could. From his own collection he contributed original old German graphics and photographs of Etruscan and Romanesque sculpture. He lent them plates made for Worringer's *Altdeutsche Buchillustration* (Old German Book Illustration), which was just then being produced. Kandinsky wrote Marc that he "would like to use them as decorations for my 'Stage Composition.' " It was Piper, too, who advised them to enrich the picture section with reproductions from Gauguin, Van Gogh, and Matisse.

Kandinsky meant to edit "a kind of almanac," and the word almanac was at first used frequently by the friends. The final design of the title page still bears this name, and—unknown until now—originally the wood block did, too. In the final printed form this word is missing. Reinhard Piper had opposed it violently. In a meeting on September 21, 1911, he won over Kandinsky, who then cut the word "almanac" out of the wood block.

But Reinhard Piper also thought it necessary to put a damper on the two friends' enthusiasm on more important questions. Their teamwork could not always have been easy, for the *Blaue Reiter* artists tended to give full rein to their horses. A glimpse into this situation is provided in a letter to Franz Marc from the publisher dated December 9, 1912, accompanying the first settlement of accounts. Since the cost was much higher than expected, Piper refers to the circumstances:

As for the production of the book, you and Mr. Kandinsky acted entirely independently of us. You never asked for our advice about accepting articles, the number of articles, or the size of the plates. You just ordered us to print or to make plates. You also fixed the retail price at ten marks without considering the cost of production.

It was correct of you to assume that a promotional pamphlet should
not be too expensive. . . . But now you must not be surprised when
despite rather active sales the cost of the book is out of proportion
to the money received. . . . You did not respect our calculation
whatsoever, and we on our part could not influence your editorial
work.

The editors' "dictatorial" manner toward their publisher is not
quite as incomprehensible as it may seem. Since a financial success
was more than doubtful, Piper safeguarded his young company against
financial loss by signing only a kind of commission contract with the
two painters. He felt first of all compelled to ask for a guaranteed sum
of money amounting to the cost of production based on a preliminary
estimate. The publisher was willing to take the moral risk by lending
the name of his company; the financial risk he had to leave to the
editors. They therefore looked for somebody to sign the guarantee.
Their correspondence reflects these endeavors in detail.

The following sentence is from Section 2 of the contract of September 28, 1911: "Messrs. Franz Marc and W. Kandinsky are jointly
liable for the covering of the cost." In a rider the publisher, on the
other hand, agrees "to pay over to the editors half of the retail price
of each copy sold." At that time the two friends had given their
signatures trusting the power of their idea. They did not know of any
realistic possibility of their covering the cost. But the following weeks
they spent reflecting. Their main hope rested again on a recommendation by Hugo von Tschudi. But Tschudi had again become severely ill.
On November 6, Marc reported on a depressing visit with him and
abruptly suggested to Kandinsky:

> . . . I am secretly convinced that if we cannot find somebody soon
> to sign the guarantee we should have the courage to get the *Blaue
> Reiter* going on our own. Without Tschudi they will not be found so
> easily. . . . Men like Koehler I would rather approach later if things
> should go wrong. First of all I am thinking of Flechtheim and
> Osthaus. I do not trust people from Munich without Tschudi
> intervening directly.

But shortly before Christmas they could report to the publisher
"that the guarantee for the *Blaue Reiter* edition as stipulated in the
contract and amounting to 3000 marks was covered. . . ."[14]

[14]The three patrons whom Marc mentioned by name in his letter had meanwhile

27

It was Bernhard Koehler finally who had agreed to cover all the expenses. Overjoyed, Marc wrote about a visit with him in Berlin on December 23, 1912: ". . . I cannot hide my delight that this one man differs in a very unusual way from all other people with whom I have had financial or some other kind of business contact during my whole life. . . ."

Wherever the name *Der Blaue Reiter* is mentioned, posterity should honor Bernhard Koehler.[15] This rich Berlin manufacturer was the uncle of August Macke's wife, Elisabeth. Even before Macke's marriage, he had supported the young artist and made it possible for him to study in Paris in 1907. At that time Koehler was already in his sixties (his son, also named Bernhard, was about the same age as Marc and Macke and was a good friend of theirs). At first "Uncle Bernhard" collected without a plan: boxes, porcelain, embroideries, also Jugendstil paintings of the Munich Scholle group. "Some good things, some not so good," Macke judged. The young painter considered it a wonderful task to awaken an understanding for modern art in the old gentleman. During a joint visit to Paris in 1908, Koehler purchased paintings of the French masters Courbet, Manet, Monet, Pissarro, and Seurat, who at that time were by no means well known in Germany. From then on Koehler's collection expanded steadily. In January 1910 he met Marc. First he aided him by buying a few of his paintings, and then, from the middle of the year on, with a monthly sum of two hundred marks; he thereby gradually accumulated the largest number of paintings by this master. This was followed by the purchase of works by other artists of this group; he bought paintings

been asked for support. After his return from Sindelsdorf, August Macke had approached the Düsseldorf art dealer Alfred Flechtheim, sending him a manuscript fragment of the planned book that was enriched with cut-out and tacked-on proofs (Documents, p. 248); the important patron Karl Ernst Osthaus in Hagen had obviously been informed in a letter from Marc. Both agreed to give five hundred marks if necessary. The editors in exchange permitted each patron to select one painting from each artist. The main part of the guarantee, two thousand marks, was assumed by Bernhard Koehler. For a whole year negotiations dragged on about running costs and proceeds. They soon broke with Flechtheim, who must have come up with commercial suggestions. His place was taken by Marc's father-in-law, who pledged five hundred marks. Finally they decided not to rely on Osthaus either.

[15]On Bernhard Koehler cf. Elisabeth Erdmann-Macke (n. 11), and Gustav Vriesen, *August Macke* (2nd ed.; Stuttgart, 1957). Also cf. Macke and Marc, *Briefwechsel* (alphabetical index), and the catalogue of the exhibition of the Bernhard Koehler donation, Städtische Galerie (Munich, 1965).

by Kandinsky, Münter, Campendonk, and Delaunay as early as the opening day of the "First Exhibition" of the editors. (He and his son had personally helped to hang the pictures for the exhibition.) Six works from his collection are included in the almanac, and they provide a survey of his far-reaching interests, showing also the assurance of his artistic taste: older art is represented by a wooden Gothic sculpture from the Rhine and an old Spanish *Death of the Virgin Mary*; newer French art by a *Half Nude* by Girieud and a still life by Cézanne. *St. John* by El Greco—which was purchased through Macke's intervention early in 1911—is boldly contrasted with *The Eiffel Tower* by Delaunay, which had been exhibited at Thannhauser's. Marc's *Bull*, with which Kandinsky later demonstrated "the powerful abstract sound of the physical form," later also went into Koehler's collection.

When the almanac was delivered, in May 1912, Marc sent the first copy to Bernhard Koehler. And many years later Kandinsky still gratefully remembered this magnanimous sponsor: "Without his helping hand, the *Blaue Reiter* would have remained a beautiful utopia."

## Plans for a Second Volume and Further Ideas

Kandinsky planned, as we have seen, a "yearbook," and in the contract with the publisher a "periodical publication" is mentioned; whenever the friends refer to their publication in correspondence, they always call it the "first volume" or "first book." But soon they realize that annual publication of a new volume was beyond their united powers. Therefore, the later press release mentioned only a publication appearing "at irregular intervals." Other decisive reasons for this were the negative experiences resulting from the pressures generated by the contemporary art scene. On May 14, 1912, Kandinsky wrote:

> The second volume: I wish we did not need to rush it at all. Let's quietly receive the material (that is, let's collect it quietly and then select more strictly than the first time). I also have a "*selfish*" wish: to have my own ideas mature for some time (at least for the summer). I am quite off the track.

But at an early stage, preparations for a second volume already ran parallel with editing work on the first almanac. From the sources we can get a very good idea of their plans for it. On the one hand, there was material for the second book from contributions that had been delivered too late or could not be included in the first.[16] On the other hand, themes already covered were to be continued. Thus Kandinsky asked his friend, who stayed in Berlin over the New Year of 1912, to look around in the Völkerkundemuseum. "Maybe we can use something later." Marc also made his own discoveries in the capital. At a dealer's he found "a huge collection of panel paintings from Athos," whose monasteries were familiar to him from a trip he had made in 1906 with his brother Paul, a Byzantine scholar. This find was, therefore, especially interesting to him! ". . . I will reproduce some of it for the *Blaue Reiter*, I am thinking of fragments for the second volume." On his way back from Berlin, Marc visited the Egyptologist Professor Paul Kahle, in Halle, and asked him to contribute an article to supplement the figures from shadow plays reproduced in the first volume. And in February 1912 the name of a scholar whose ideas came closest to those of the *Blaue Reiter* was mentioned: "I am just reading"—Marc wrote to Kandinsky—"Worringer's *Abstraktion und Einfühlung* [*Abstraction and Empathy*], a good mind, whom we need very much. Marvelously disciplined thinking, concise and cool, extremely cool."

The inclusion of authors such as Carl Einstein, Kahle, and Worringer indicated the greatest step beyond the previous program, which had followed the—by the way rather disputed—maxim of Delacroix to let only artists speak. Later on Kandinsky often gave reasons for this new attitude. In 1935 he wrote: "For the second volume of the *B.R.* we intended to draw upon scholars as collaborators in order to expand the earlier basis of art and in order to show how the work of the artist and the scholar is related and how close together their two spiritual fields are."

At that time August Macke too was greatly interested in plans for further publications. Among Franz Marc's papers, a letter from Macke to Kandinsky was found, dated July 30, 1912. He mentions a most

---

[16]This is true for P. P. Girieud's contribution on "Siena" listed in the "provisional table of contents," for N. Brüssov's article "On Musicology" announced in the subscription prospectus, and for an essay by the art critic Carl Einstein.

significant meeting with Theodor Däubler, the patriarch of expressionism. He calls the poet:

> . . . a most outstanding man. I am sure he will visit you in Munich.
> He is three times as tall and three times as heavy as I am. . . . He
> is as overflowing in his poems as you are in your paintings. I asked
> him whether he would possibly collaborate on the second volume
> of the *Blaue Reiter*. He agreed gladly for he values the *Blaue Reiter*
> highly.

By the middle of 1913 continuation of the enterprise was still
unquestioned. A note from Kandinsky on June 5 sheds an interesting
light on the state of preparations:

> . . . I believe that we will scarcely be able to come out with the
> second volume next winter. Where can you find the material,
> especially good articles? Up till now I have received offers from
> Busse, Reuber (Berlin), Larionov. I asked them all to send manu-
> scripts but did not promise anything. Wolfskehl would like to write
> something, and we can be sure that it won't be bad. I also have two
> prospective Russians. Both, I think, can write something good.
> Whether they will do it and when remains to be seen. And pictures?
> Up until now I have one idea only, which for the time being I ask
> you to keep absolutely secret, your wife, of course, excepted. They
> are old signboards and advertising displays, in which I include
> paintings on concession stands (e.g., Oktoberfest fairgrounds).
> I intend to go to the border of kitsch (or as many will think, *across*
> the border). In connection with this, nature photographs, especially
> single objects and parts of them. Etc. . . .

Again it is amazing how much the Russians were supposed to
contribute. Recently it has been stated that Larionov and Goncharova
had already separated from the "Munich decadence" in 1912;[17] this
can be refuted by pointing to the fact that both participated in Walden's
Erster deutscher Herbstsalon (First German Fall Salon) the following
year. Kandinsky's letter finally further confirms this: Larionov offered
to collaborate with the *Blaue Reiter* as late as the middle of 1913. We
may assume that for the second volume he would have written on
"rayonnism," which he had just propagated in a manifesto. Another

---

[17]Camilla Gray, *The Great Experiment: Russian Art 1863-1922* (London and New
York, 1962), p. 121.

dimension of the new volume is suggested by the mention of Karl Wolfskehl (1869–1948). This is not the place to trace the extensive significance of this man and of the Schwabing "Cosmic Circle" in the cultural atmosphere of Munich. Wolfskehl was a friend of Kandinsky's as well as of Marc's. He had contributed to the *Jahrbuch für die geistige Bewegung* (Yearbook of the Spiritual Movement), edited by Friedrich Gundolf and Friedrich Wolters, and this had long ago introduced him to the *Blaue Reiter*, for the ideas of the two groups were surprisingly close. And, finally, Kandinsky's idea of discussing kitsch, documented so originally by company and booth signs, could have been most important. It shows an unmistakable perception of a problem of the future.

In the meantime a new project was being pursued. Elsewhere I have described extensively a plan to illustrate the Bible, which dated back to the spring of 1913 and was to be executed jointly by Marc, Kandinsky, Kubin, Klee, Heckel, and Kokoschka. For our present purposes it is interesting to note that they announced this publication as a *Blaue Reiter* project. This again proves that there was no fixed society or group by that name, but that Kandinsky and Marc as "editors" always "freely selected" the collaborators. As a result of the war, the illustrated Bible remained unfinished, but a considerable number of works can be attributed to the joint enterprise.[18]

In 1913 it became obvious that a second edition would soon be necessary. True, the first printing had been increased to twelve hundred copies as more people than anticipated had subscribed. But their number was still not large. The publisher had cleverly kept the type standing, so changes were, of course, impossible. Half a sheet, however, was added in front for the two new prefaces by Kandinsky and Marc (Documents, p. 257). And, what was most important, a suitable place could now be found on a right-hand page for the dedication to Tschudi. As may be remembered, the two editors had been very angry about the unsatisfactory solution of placing this dedication on the verso of the last page. Also, a new plate was made for the Rousseau self-portrait, and Marc replaced the color plate of *Horses* with another. The prefaces can be dated March 1914; the delivery of the volume must, therefore, have occurred in early summer.

---

[18]Klaus Lankheit, "Bibel-Illustrationen des Blauen Reiters," *Anzeigen des Germanischen Nationalmuseums: Ludwig Grote zum 70. Geburtstag* (Nuremberg, 1963), pp. 199ff.

Preparations for another volume of the almanac were going on at **History**
the same time. Around the turn of 1913-14, the two friends had the
idea of alternately supervising a volume so that the other would have
more time for his own artistic work. Impatiently Marc wrote the
preface to the next volume (Documents, p. 260). Kandinsky became
acquainted with the Serbian writer Dimitri Mitrinović and wrote to
Marc on February 17, 1914: "He can be *very* useful to the *Blaue
Reiter*." Even in the preface to the second edition, dated "March
1914," Kandinsky emphasized the "necessity for a further development
of these ideas" (Documents, p. 258). But soon his apprehension "that
times were not ready for the *B.R.*" increased; he again proposed to
delay the second volume. Primarily committed to his artistic work
he had to admit that he was waking up at night thinking of nothing
but the *Blaue Reiter* (March 10, 1914). Disregarding this, Marc, the
fighter, felt compelled to speak out:

... Up to this day I still feel the drive or rather the urge to increase
and to clarify my present joy of working through written editions.
Maybe I will try something on my own. I feel that especially now
and especially because there is no material, we must say something.
When it is no longer lacking, others will provide the material with
more justification. . . . (March 13, 1914)

The Munich movement found a new herald in Hugo Ball, theatrical
producer at the Kammerspiele. As is obvious from his diaries, he
venerated Kandinsky as a "prophet of renaissance." During that same
March 1914, he developed his revolutionary ideas about the "expressionist theater." Soon after, Ball joyfully wrote to his sister:

As far as this new idea is concerned, I am planning a new book for
the first of October, *The New Theater*, together with Kandinsky,
Marc, Thomas von Hartmann, Fokine, von Bechtejeff. We jointly
want to develop our ideas about the new artistic theater with new
scene paintings, music, figurines, etc. We would like to meet in
June at Lake Kochel in order to organize everything. Also new
architectural plans should be made. A completely new theater.
A new festival house. . . . If we succeed in bringing out the brochure
before October 1 (Piper, Munich) [we will found] an "International
Society for Modern Art," including not only the theater, but also
modern painting, modern music, modern dance. (May 27, 1914)

33

In his diaries Ball again mentioned these joint plans; he also gave an outline of the contents:

My thesis is that the expressionist theater is an idea for festivals, it includes a new concept of the total work of art. . . . Kandinsky introduced me to Thomas von Hartmann. He had just returned from Moscow and had a great deal of news about Stanislavski: how, influenced by Indian studies, one gives performances of Andreyev and Chekhov there. This was different, broader, deeper than we were, also newer. And it contributed much to broaden my view and my demands for a modern theater.

Theoretically, the artistic theater should have about the following structure:

| | |
|---|---|
| Kandinsky .   .   . | Total work of art |
| Marc .   .   . | Scenes for *Storm* |
| Fokine   .   . | On Ballet |
| Hartmann .   . | Anarchy of Music |
| Paul Klee .   . | Designs for *The Bacchantes* |
| Kokoschka   . | Scenes and Plays |
| Ball .   .   . | Expressionism and the Stage |
| Yevrenov   .   . | On the Psychological Element |
| Mendelsohn   . | Stage Architecture |
| Kubin   .   . | Design for *The Flea in the Fortified House* |

Carl Einstein's *Dilettantes of the Miracle* indicated our direction.[19]

It is hardly necessary to explain that the ideas outlined by Ball were almost identical with those of the *Blaue Reiter*. Congenial minds had met. To our surprise we recognize an overwhelming majority of collaborators from the almanac; we see also that the *Blaue Reiter* editors were put in first and second place. Hartmann, Kokoschka, and Klee were very close friends of Kandinsky's and Marc's; they also had long been acquainted with Fokine. The mention of Einstein, whose collaboration had already been foreseen for the second volume of the almanac, underlines this relationship. With the architect Mendelsohn entering the scene, the basis was perforce widened to include architecture. Even the publisher was to remain the same for this volume.

[19]Hugo Ball, *Briefe 1911–1927* (Einsiedeln, Zurich, Cologne, 1957), pp. 29f.; Hugo Ball, *Die Flucht aus der Zeit* (Lucerne, 1946), pp. 8, 10, 12f. (English translation to be published by The Viking Press, New York, 1974, in The Documents of 20th-Century Art.)

Kandinsky later blamed only the war for interrupting the collective work; he said as much in his note of 1935. But when in the 1920s the publisher suggested a new edition, Kandinsky—according to his wife —responded more than once, "The *Blaue Reiter*—that was the two of us: Franz Marc and myself. My friend is dead, and I do not want to continue alone." This noble attitude we find again in an unpublished letter to Sir Herbert Read of November 18, 1933, saying: ". . . I declined a third [edition] because Marc was dead."

It is fruitless to ask whether a further development would have been possible. A fortunate meeting of true historical significance, like that of these two congenial artists, is always a short, propitious moment and it shines far into the future.

## Content and Effect

The spontaneous origin and the fragmentary shape of the book make its success seem all the more remarkable. Yet we must warn the reader not to study it as a systematic compendium of contemporary art theory. Its effectiveness as a torch is due to the persuasive power of a few simple truths that were proclaimed at the right time. The book rests on two cornerstones: the three short, aphoristic introductory articles by Marc and the three more comprehensive concluding contributions by Kandinsky. Between them are the articles by the collaborators and almost one hundred and fifty interspersed illustrations. The book presents the vision of an ideal future empire of art rather than a scientific system. One may even discover in it some utopian aspects but Kandinsky, Marc, and their companions did nothing less than outline a program of modern aesthetics and articulate principles of artistic creativity that are still of current interest fifty years later.

Whenever we call the *Blaue Reiter* artists "revolutionaries," we do not mean that they denied all tradition. One of their prime concerns was to trace back to the true sources of art. According to Kandinsky's plan, a "link to the past" was a prerequisite for a "ray to the future," which they were hoping to achieve. Already the term *Blaue Reiter* sounds strange but not really peculiar. The casual description of that afternoon in Sindelsdorf says nothing about the spiritual background, which alone would shape the name. It even seems as if the artist intentionally wanted to avoid impertinent questions. He more or less took the explanation of the name for granted. And, in fact, however many derivations and associations may suggest themselves, we all seem to understand fully what Kandinsky and Marc meant. The coordinates of this historic point of encounter are: the "rider," indicating noble-mindedness to every European; and the color blue, indicating depth of feeling ever since the romantic movement with its longing for spiritual fulfillment. The final cover design created for the almanac by Kandinsky after numerous sketches, still gives a sense of this Western-Christian tradition, for the figure of the Blaue Reiter is modeled after the traditional holy figures of the knights St. George and St. Martin. The composition in two colors, which exist independently of each other but still form a unity, is like a symbol of this creative friendship.[20]

To trace the many-faceted presuppositions for the idea-world of the *Blaue Reiter* would call for nothing less than an outline of European cultural history of the nineteenth and the early twentieth centuries. For the roots go back to the classical-romantic period. Threads could be spun from French symbolism and from Russian mysticism, which both go back—via Schelling and Hegel—to the German romantic movement. Wagner and Nietzsche left their marks, as did the aesthetics of the Jugendstil and neo-idealism. Exciting parallels suggest themselves, such as Freud's depth psychology, Husserl's principle of phenomenology and especially Bergson's philosophy of the *élan vital*. Bergson's *Introduction to Metaphysics* appeared in the same year, 1912, as Kandinsky's and Marc's almanac and—as Ernst Troeltsch has shown—it too had been strongly influenced by the classical and romantic movements. We have already pointed out how surprisingly close Worringer's *Abstraction and Empathy* (published by Piper in 1908) is to the ideas of the *Blaue Reiter*.

All voices of the time had something in common: a rejection of the claim of the infallibility of empirical knowledge. In his treatise Kandinsky judged that the "spiritual crisis" was spreading so stormily that even the sciences were being torn apart and we were standing "in front of the door which leads to the dissolution of matter." The rhapsodic and imploring tone that dominates the *Blaue Reiter* can be understood only against the apocalyptic enthusiasm of the eve of World War I: "Kandinsky maintains that the end of the

[20]A red plate was probably used later and only for the cloth edition.

36

nineteenth century and the beginning of the twentieth is the start of one of the greatest epochs of man's spiritual life. He calls it 'the epoch of great spirituality.' " Thus the artist viewed himself in a "Self-Characterization" as late as 1918–19. In the *Blaue Reiter* Franz Marc said he was convinced "that we are standing today at the turning point of two long epochs, similar to the state of the world fifteen hundred years ago" (see p.[12]).

It was at that time widely held that a new spiritual epoch would take the place of nineteenth-century materialism and that it was art that was called upon to introduce this turning point. Hugo Ball outlined the objective, of which he considered Kandinsky the "prophet": "The renascence of society out of a union of all artistic means and powers." This search for a synthesis of culture is also a predominant characteristic of the *Blaue Reiter*. The old German romantic idea of a total work of art had entered a new stage toward its realization. These artists were always concerned with more than just "art." Marc demanded "that the renewal should not be simply formal but a rebirth of thinking." Aware of this mission, he defined the task of the artists: "To create out of their work *symbols* for their own time [which should be placed] on the altars of a future spiritual religion, symbols behind which the technical heritage cannot be seen" (p.[7]). Kandinsky, too, clearly confessed a social responsibility, and he deplored the "total loss of a mutual relationship between art and human society." As late as 1936, he described his own mission as follows:

> Marc and I had thrown ourselves into painting, but painting alone did not satisfy us. Then I had the idea of doing a "synthesized" book which was to eliminate old narrow ideas and tear down the walls between the arts . . . and which was to demonstrate eventually that the question of art is not a question of form but one of artistic content.

Consequences for art were drawn very clearly from this situation. It was Franz Marc, who, initially referring to El Greco and Cézanne, found a poetic phrase describing the task of the artist in our time, which has justly become famous: "the *mystical inner construction*, which is the great problem of our generation" (p. [3]). In his essay on cubism, Roger Allard used the phrase "dematerialized their conception of the world" (p. [38]). August Macke defined form as a "mystery . . . for it is the expression of mysterious powers" (p.[21]). In his treatise (*Con-*

*cerning the Spiritual in Art*) Kandinsky had already mentioned a "prophetic word of Goethe's"; and now in the almanac he quoted a passage from Riemer's *Gespräche mit Goethe* (Conversations with Goethe), which may be considered a leitmotiv for all their joint endeavors: "In 1807 Goethe said: . . . 'in painting the knowledge of the thorough bass has been missing for a long time; a recognized theory of painting, as it exists in music, is lacking' " (p. [42]).

The longest essay of the almanac was dedicated to these aspects of "the question of form" (p. [74]). Grohmann called it "Kandinsky's most mature contribution to art theory." In it the artist propounds the principle of "inner necessity," which he has always considered the center of his convictions. It allowed an unrestricted external freedom in the choice of means of expression and it permitted an art that encompassed the two poles of "total abstraction" and "total realism." It also justified the initially confusing number of illustrations in the book; in an announcement supplementing Kandinsky's treatise it is explained that "the publication is a meeting place for all powerful endeavors that can be observed today in all fields of art; their basic tendency is to expand the former limits of artistic expression."

The reproductions were considered an essential part of the book. Since it was not the "conventional exterior" but the "inner life" that determined the value of a work of art, all standards of classical aesthetics were rendered invalid (p. [10]). "With a divining rod we searched through the art of the past and present," they said in the Preface to the Second Edition (Documents, p. 258). The pictures were carefully placed in relation to each other so that the "inner sound" of one picture would be answered by the "contrasting sound" of the other. At the same time the examples of older proved cultures caused an "ordeal by fire" for their own endeavors (p. [8]). Thus Kandinsky anticipated a valuable experience for the reader who would approach the pictures with the right attitude:

> If the reader is able to rid himself of his own desires, his own ideas, his own feelings for a while and leafs through this book, going from a votive painting to Delaunay, from Cézanne to a work of Russian folk art, from a mask to Picasso, from a glass painting to Kubin, etc., etc., then his soul will experience many vibrations and he will enter the sphere of art. (p. [99])

Expansion of the traditional boundaries of artistic expression was put forth as a "basic tendency" of the editors as we have outlined

38

above in context with the book's genesis. It was to a great extent
owing to this provocative juxtaposition of works of pictorial art from
different periods and different cultures that the *Blaue Reiter* gave the
appearance of being revolutionary.[21] But the trend had precedents.
During the romantic period old German graphics and medieval or
Persian miniatures were rediscovered; since the second half of the
nineteenth century, "l'art japonais" or "l'art primitif et l'art populaire"
had become stimulating phrases for the younger French generation.
Finally Picasso and the painters of the Brücke had, independently of
one another, hailed African art and other treasures from ethnographic
museums as an unexpected confirmation of their own "urge for
abstraction." But it was the *Blaue Reiter* that brought about the great
synopsis of expressions from previous epochs and foreign cultures
that were diametrically opposed to the academic naturalism of
established art. Along with this there came independent discoveries.

The figures from Egyptian shadow plays were discovered almost
accidentally. Scholars had just recently presented them to academic
circles. Dr. Paul Marc had called Kandinsky's attention to the magazine *Islam*; originally Kandinsky had intended to use the illustrations
from Professor Kahle's article for a projected theory of composition.

The Nabis already had prized the *Images d'Epinal*, examples of
which Girieud was asked to contribute. The Russian folk prints—
the *lubki*—may be considered an Eastern parallel. Their naive anti-
naturalistic style corresponded very closely to the ideas of the modern
artists. The *Blaue Reiter* introduced them for the first time to foreign
art lovers. Kandinsky added to the Goltz exhibition of graphics eight
new prints from old wood blocks that he owned—and it was this
very "mockery" that shocked the public. He had added an explanatory

---

[21]In general, their feeling for good style is so remarkable that they had no difficulty
in selecting works from different epochs that had just come to light. On the other
hand, it would seem to be a whim of history, which is not uncharacteristic of
such a stormy movement, that minor works could also be used effectively; at one
point, even a forgery fired their creative imagination. Their examples of Japanese
and Chinese art cannot all be called Oriental masterpieces, and the "Gothic"
knight with the inscription on a ribbon accompanying Goethe's programmatic
saying (p. 112), which also decorated the first page of the subscription prospectus,
was later discovered to be a nineteenth-century imitation. Cf. W. L. Schreiber,
*Handbuch der Holz- und Metallschnitte des XV. Jahrhunderts*, Vol. IV (Leipzig,
1928), no. 2107 m: "Unknown saint . . . Munich, Hahlweg & Stöckle . . . belongs
to series no. 2063 a: 'Di/Heiligen/der Landschaft/Passel/1818' " (reference
kindly given by Wolfgang Wegner).

text to the catalogue, which he transferred to the *Blaue Reiter* almost literally:

> These prints were primarily produced in Moscow during the first half of the nineteenth century (the tradition dates back to a much earlier time of course). Itinerant booksellers sold them even in the remotest villages. One can still see them in farmhouses, although they are largely being replaced by lithographs, oleographs, etc.

The first reproductions listed in the provisional contents—before the French and Russian folk prints—are "Bavarian Glass Paintings" (Documents, p. 245). Peasant glass painting influenced the views and styles of the Munich group more than any other kind of folk art.[22] Murnau, on Lake Staffel, was a center of this cottage industry; Jawlensky and Werefkin stayed there frequently, Kandinsky and Münter lived there. They collected these examples of a dying period. A visitor reports: "Jawlensky had a whole wall of his Munich studio covered with glass paintings." Gabriele Münter's biographer wrote: "Groups of those paintings bought at county fairs were hanging all over the apartment." The painter friends studied particularly the numerous outstanding examples collected by Krötz, a brewer in Murnau, and they selected no less than eleven examples from his collection for the almanac. Their publication introduced peasant glass painting into the history of art. As if this were not enough, the artists, inspired by this old means of cultural expression, were stimulated to try paintings of their own in this medium. In creative imitation, nearly all of them made glass paintings. And even the symbol—the figure of the blue rider—owes its final character to these intensive studies of folk art. After many different sketches, Kandinsky decided to use a rider from one of his own glass paintings for the cover.[23]

For Christmas 1911 Marc gave his friend a small painting he had done in this technique. It shows the painter Henri Rousseau. Reinhard Piper had sent Kandinsky a copy of a small Rousseau monograph by

[22]Klaus Lankheit, *Hinterglasmalerei im XX. Jahrhundert*, exhibition catalogue (Mainz, 1962).
[23]Eberhard Roters, "Wassily Kandinsky und die Gestalt des Blauen Reiters," *Jahrbuch der Berliner Museen*, V (1963), 201ff. Also cf. Kenneth Lindsay, "The Genesis and Meaning of the Cover Design for the First *Blaue Reiter* Catalogue," *The Art Bulletin*, XXXV (March 1953), 47ff.; Hideho Nishida, "Genèse du Cavalier bleu," *XXe siècle*, No. 28 (December 1966), 18ff.

Wilhelm Uhde, which had just been published in France. Kandinsky already knew the works of this painter from Paris, and in his letter of June 1911 he had mentioned Rousseau. But now this *peintre naïf* came to him as a revelation. In him Kandinsky saw his own counterpart, and in the article on the question of form he wrote: "Henri Rousseau, who may be considered the father of this realism, has pointed the way simply and convincingly" (p. [94]).

He wrote to Uhde asking him to lend him a few plates for the *Blaue Reiter*, and through the intervention of Delaunay he purchased *The Street* from Rousseau's estate (illustration, p. [89]). Although the paintings were reproduced rather poorly in the monograph, Marc became so enthusiastic that he copied the self-portrait of the *Douanier* in glass-painting technique and presented it to Kandinsky at Christmas. It was a feat and also a risk to publish six examples of this unknown "primitive painting," which were thus elevated to works of art! At that time Reinhard Piper did not dare publish a German edition of Uhde's monograph as Delaunay had advised him to. Alfred Flechtheim of Düsseldorf finally published it in 1914.

Today we take this kind of painting for granted; we are also interested in the art of children. But we must not forget that here, too, it was the *Blaue Reiter* that paved the way and awakened understanding. Not only does the almanac contain a series of children's drawings but—in his basic article on the question of form and in his reference to the amateur painting of a Rousseau—Kandinsky also tried, using psychological reasoning, to teach people to understand these creative exercises: "The child is indifferent to practical meanings since he looks at everything with fresh eyes, and he still has the natural ability to absorb the thing as such. . . . Without exception, in each child's drawing the inner sound of the subject is revealed automatically" (p. [92]).

This idea of the "inner sound," which the painter senses in each object and which he reproduces for the viewer, provides the easiest access to the idea of analogy or synesthesia of the arts, and to the demand for their "synthesis." This is exemplified by the relationship of painting and music, to which much space is devoted in the almanac. The idea is not new; Plato had said: "Painting is controlled by the same laws as musical rhythm" (*Republic* III, 10 ff.). Since the preromantic movement the idea of "color-music" had become topical, and, by about 1900, the idea of a relationship between these arts may be considered a general conviction of painters and musicians. But here again the *Blaue Reiter* was the first to be serious about its realization. Kandinsky was the very man to bring this into effect, for at first he intended to give his *Concerning the Spiritual in Art* the subtitle "Language of Color." On the one hand he had eagerly studied Russian efforts, as he pointed out in an appendix to his treatise, and it was therefore natural for Kandinsky to ask Moussorgsky's disciple, Leonid Sabaneiev, to describe Scriabin's color symphony *Prometheus: The Poem of Fire*, which "is based on the principle of corresponding sounds and colors" (p. [59]). On the other hand, his friendship with Arnold Schönberg—"one of the most amazing conjunctions in the firmament of the twentieth century" (H. H. Stuckenschmidt)—pointed to a definite possibility for collaboration. The composer's *Theory of Harmony* was to appear at Christmas in 1911, at the same time as the painter's *Concerning the Spiritual in Art*; this was not only superficially coincidental, for both turned out to have the same basic attitude: "The artist creates not what others think is beautiful but whatever is necessary for him." As Kandinsky had written to Marc: "Schönberg *must* write on German music," he was just as determined when he asked the musician to collaborate. "The first issue without Schönberg! No, this is impossible" (September 16, 1911). In fact, Schönberg's contribution, "The Relationship to the Text," reveals amazingly similar ideas. Kandinsky did not begin with theoretical principles but with visual emotions, and Schönberg likewise begins with an experiment in sound. The words from the final passage could have been written by Kandinsky: "An apparent divergence on the surface can be necessary because of a parallel movement on a higher level" (p. [33]).

In 1909 Kandinsky had met the Russian composer Thomas von Hartmann in Munich. It was another fortunate coincidence that Hartmann returned to Germany and that he and Kandinsky could exchange ideas intensively. "On Anarchy in Music" is the result of their talks. Hartmann also refers to the concept of "inner necessity" and defines "the essence of beauty in a work" as "the correspondence of the means of expression with inner necessity" (p. [43]). In an unpublished speech in New York in 1950, Hartmann reminisced about his friendship with the painter and called Kandinsky's composition "The Yellow Sound" (p. [115]) "the greatest venture of stage art to this day." One should not study this design for an abstract, total work of art (which according to Grohmann had been written out already in 1909), without reading the author's introductory principles, "On Stage Composition" (p. [103]). Only then will the reader appreciate it as a logical continuation of his ideas on a "synthesis" of the arts and also as a glorious finale for the whole almanac. Nowhere else in the book is it as clear how deeply the *Blaue Reiter* is rooted in the art theory of the pre-romantics: Philipp Otto Runge, the friend of Novalis, had planned "an abstract, pictorial, fantastic, musical work of poetry with choruses."[24]

Just as Runge's dream remained unfulfilled, so, too, did Kandinsky's vision of a century later. Kandinsky remarked in a footnote that the "Musical portion [was] by Thomas von Hartmann" (p. [117]). Hartmann confirmed this in his New York speech, and added he had presented the work to the Moscow Art Theater together with his music and designs for the setting by Kandinsky: ". . . but they could not understand it and did not accept it. The designs and my music—everything was lost during the Revolution." Only in 1956—almost a half-century later—did Jacques Polieri and Richard Mortensen take up the work again.

Following that "link to the past" we have also unexpectedly seen some of the "ray to the future." We would necessarily go wrong if we tried to ascertain the actual impact of the almanac in detail. It would be impossible because we could never prove exactly how many and what kind of contemporary readers the book had who were stimulated by word and picture. Also, the book was only one means of acquainting the public with the *Blaue Reiter* movement. And it signified only one—though probably the most important—stage in the development of Kandinsky, Marc, and some of their friends. On the other hand, August Macke's abundant paintings, as his biographer has rightly emphasized, are not simply the result of this contact, but must be evaluated as an individual contribution to twentieth-century painting. By the same token, the ideas of Paul Klee, who is represented in the only published volume by only one little drawing, should be taken into consideration; he is as important for an evaluation of the effects of the circle as is Delaunay's orphism. In other words, the history of the *Blaue Reiter* does not end with the history of the book that bears this name. Here, however, we are concerned only with the latter.

Instead of using platitudes, let us rather look for reliable indications that might show the degree of understanding—positive and negative—

---

[24] Klaus Lankheit, "Die Frühromantik und die Grundlagen der 'gegenstandslosen Malerei,'" *Heidelberger Jahrbücher* (1951), pp. 55ff., especially p. 68.

which followed the book's publication. Significantly enough, it was an art historian of the Viennese school, Hans Tietze, a man open to new trends, who, in my view, wrote the most sensible and comprehensive critique: his was one of the very few positive reviews, a fact that in itself makes its preservation worthwhile. Tietze announced the book in the magazine *Die Kunst für Alle* (Art for Everybody).[25] After summarizing the contributions of the book, Tietze discusses these works of art and the theses of their authors in their historical context. He understands this art as "an intentional reaction against impressionism" and "against an illusionistic reproduction of nature." Taking up Roger Allard's term, he says: "The new style dematerializes the conception of the world." As to their attempts to create an artistic language "of lines and masses, of colors and shades, which are not meant to remind one of definite objects or to include intellectual associations," he remarks, correctly, that the movement was "still too much in the process of development to have already produced a complete grammar of forms, a thorough bass, as the *Blaue Reiter* says, borrowing a word from Goethe." The critic thinks it characteristic that "this movement should bloom at a time when in aesthetics too expression, as a basic element of art, is emphasized more than heretofore." He warns his readers not to identify this movement with a theory, which would be as lifeless as all theories, and he admits that because of their forms and colors many new works caused an echo "of a wonderful and almost incomprehensible intensity." He is convinced that "this struggle for the spiritual, this fight for a new synthesis has not been in vain: We are reminded with a rare insistence that the imitation of nature and the representation of reality are not the goals of art."

Finally, Tietze confirms a harmony with true tradition and celebrates the publication of the book

> as a courageous and liberating achievement. At this time there is nothing more beneficial to German, indeed all art, than this moving memento with its inexorable consequences; it reminds us of the wonderful privilege that Albrecht Dürer claimed for the artist: to create a world out of himself which outside himself would not exist.

The relationship of the *Blaue Reiter* to contemporary aesthetics

[25]Hans Tietze, "Der Blaue Reiter," *Die Kunst für Alle*, XXVII (1911–12), 543ff

raises the question of whether the book could also have enriched the methods of art theory. As early as 1914 Heinrich Wölfflin propounded the thesis that "history of art and art run parallel." A few examples will demonstrate that the question is at least justified. When we discussed the "comparative history of art" as it is demonstrated in the *Blaue Reiter*, the idea struck us that the juxtaposition of pictures from different epochs occurred here for the first time; this would provide us with an important art-historical method. Furthermore, Ludwig Grote has pointed out that the interpretation of Gothic art in general was influenced by Delaunay's *St. Séverin*, published in the *Blaue Reiter*.[26] And when, in his inaugural lecture at the Technische Hochschule of Stuttgart in 1912, Hans Hildebrandt attempted to prove that it was entirely possible and legitimate to have paintings without concrete contents, this, too, may be regarded as a result of the book and the treatise by Kandinsky, published a short time before.[27] It is doubtful whether Max Picard's book *Expressionistische Bauernmalerei* (Expressionist Peasant Painting) of 1918, could have been published without the previous publication of peasant glass paintings in the *Blaue Reiter*. Finally, an impartial witness is again Heinrich Wölfflin. With him, as we all know, the comparative interpretation of works of art reached its pinnacle. Referring to the Wölfflin edition of the *Bamberg Apocalypse* of 1918, Hans Jantzen recently pointed to "the sudden reference of the author of *The Art of the Italian Renaissance* to a manuscript that can no longer be interpreted in terms of Renaissance art." Actually the scholar's statement touches not only on Worringer's treatise but also on the almanac: "In a conspicuous extension of certain trends in modern painting, we have learned only recently to notice positive qualities in the so-called stiffness, to accept a different intention instead of criticizing it as a lesser quality."[28]

The "powerful, catastrophic shock of our whole culture," which Tietze considered the prerequisite for understanding modern art, was to follow all too soon. But even World War I could not extinguish the flame that Kandinsky and Marc had kindled. From Munich their ideas spread to other centers. Herwarth Walden's *Sturm* had actively stood up for the *Blaue Reiter* group as early as 1912. Through Kubin,

[26] *Der Blaue Reiter*, exhibition catalogue (Munich, 1949), p. 12.
[27] Unpublished manuscript, kindly made available to the author by Frau Lily Hildebrandt.
[28] Hans Jantzen, "Heinrich Wölfflin," *Geist und Gestalt* (Munich, 1959), pp. 297f.

45

Lyonel Feininger had joined them in the following year; Franz Marc's letter inviting him to collaborate is decorated with the *Blaue Reiter* symbol. Both during the war and afterward Walden, through his magazine and his exhibitions, did, though in a rather self-willed way, play an important role in bringing this art to the fore.

Through the *Sturm* connections were made with Zurich, where a new group of emigrants had gathered. Hans Arp had already participated in the graphics exhibition in Munich and had been represented in the almanac with a drawing and several vignettes. Then Hugo Ball came to Switzerland. A note in his diary in 1917 reveals his spiritual heritage: "Yesterday my lecture on Kandinsky. An old favorite idea of mine came true. Total art: pictures, music, dance, poetry—everything is there. . . ."

The "Program of the Second (*Sturm*) Soiree," published a few days later, contains well-known names: Kandinsky's poems "Bassoon," "Cage," "Glimpse and Flash" were read and Walden's eulogies of August Macke, Franz Marc, and August Stramm; Oskar Kokoschka's *Sphinx and Strawman* was performed.[29]

The founding of the Bauhaus, however, was much more momentous than those endeavors. It may justly be called the legitimate continuation of the *Blaue Reiter* on a new historical level. Of course, the earlier movement was not the only basis for the new foundation, and the Bauhaus did not take over the ideas of the *Blaue Reiter* unchanged. Changes within the Bauhaus itself should also be taken into consideration. But an irrefutable proof of the continuation of the pre-war movement is the fact that Walter Gropius appointed the surviving masters of that group—Kandinsky, Klee, and Feininger—to this school. As we have learned only recently, it was only tragic misunderstanding that prevented Schönberg from going to Weimar and enriching this intellectual, artistic center with music; had he done so, the continuity would be even clearer.[30] Walter Gropius's manifesto, decorated with the famous woodcut of the "Cathedral of Socialism," contains the same—I am again tempted to say utopian—character as the "visionary advent" of the *Blaue Reiter*; and down to its very diction it can be compared with Franz Marc's notes. When, moreover, Kandinsky, Klee, and Feininger became even more closely related to

[29]Ball, *Die Flucht aus der Zeit*, pp. 148, 150, 157.
[30]Arnold Schönberg, *Briefe: Ausgewählt und herausgegeben von Erwin Stein* (Mainz, 1958), letters nos. 42, 63, 64 to Kandinsky and p. 90n.

Jawlensky, they called this new association the *Blaue Vier* (Blue Four).
Klee confirmed explicitly that the name was intended to indicate the
spiritual basis of friendship of the four artists. Galka Scheyer, soul
and propagandist of the *Blaue Vier*, selected this title intentionally
referring to the *Blaue Reiter* in Munich: "They did not want to start
something new but to emphasize something already existing; they
were all striving after 'the spiritual in art,' which was not to be interpreted but proclaimed."[31]

As late as 1930 Kandinsky still mentioned the "*Blaue Reiter* idea,"
and Mies van der Rohe also called the Bauhaus "an idea." What they
had in common was the thought of a synthesis of culture encompassing
all fields. In Weimar and Dessau, just as in Munich a decade earlier,
they wanted to re-evaluate the element of structure: Kandinsky's
*Point and Line to Plane* and Klee's lecture on "Thinking in Images"
have become textbooks for the thorough bass of painting, which
Goethe had demanded; the principles they established have since
conquered the world. But just like the *Blaue Reiter*, they reached
beyond art, aspiring to a renewal of the total culture. The creators
and masters of the Bauhaus also felt morally responsible, and—in
Paul Klee's words—they struggled for "a union of philosophy of
life and pure creative art," a union that is still one of our most serious
problems.

The Bauhaus dates back for more than one generation; six decades
have elapsed since the *Blaue Reiter*. We no longer share their high-
flying optimism that we are approaching a new "epoch of the spiritual"
and that it is the artist above all who is called upon and who has
the power to save mankind from the abyss through his works. We
doubt their conviction that art could have an effect on the masses.
In the ideas of the *Blaue Reiter* we see more clearly the expression of that time. But their postulate of the subjective "freedom" of
creative man, which is objectively justified by the "inner necessity,"
should still obtain. Marc's words have never been truer "that art was
concerned with the most profound matters, that renewal must not be
merely formal but in fact a rebirth of thinking" (pp. [6–7]).

One may definitely point to the personal and contemporary circumstances of the *Blaue Reiter*. One may classify this movement as the
highlight of a specific Munich art. One may trace back this wealth of
ideas to a national heritage—German, Russian, or French. Heinrich

[31]Clemens Weiler, *Jawlensky* (Cologne, 1959), p. 119.

47

Wölfflin, the great Swiss scholar, who was appointed to the Munich chair for history of art in the very same year that the *Blaue Reiter* was published, stated correctly "that the highest values of art are not identical with mere national characteristics. It is," says Wölfflin, "a characteristic of all great art that it projects into the sphere of general humanity."

# Der *Blaue Reiter*

Bavarian Mirror Painting: St. Martin

# DER BLAUE REITER

HERAUSGEBER: KANDINSKY
FRANZ MARC

MÜNCHEN, R. PIPER & CO. VERLAG, 1912

To the memory of
*Hugo von Tschudi*

Kandinsky [only in the de luxe edition]

Franz Marc [only in the de luxe edition]

# Spiritual Treasures

## by Franz Marc

German [Fifteenth Century]

It is strange that people should value spiritual treasures so completely differently from material ones.

If, for example, someone conquers a new colony for his country, the whole country rejoices for him, and does not hesitate—even for a day —to take possession of that colony. Technological achievements are welcomed with the same rejoicing.

On the other hand, if someone should think of giving his country a new purely spiritual treasure, it is almost always rejected with anger and irritation; his gift arouses suspicion, and, people try, in every possible way, to do away with it. If it were permitted, the donor would, even today, be burned at the stake for his gift.

Isn't this a horrible fact?

A small, recent example led us to make these opening remarks.

Meier-Graefe[1] undertook to present his countrymen with the marvelous ideas of a great master completely unknown to them— we are speaking of El Greco. The general public, even the artists, not only remained indifferent, but attacked him with undisguised hostility and indignation. This simple and noble act made his position in Germany almost untenable.

It is terribly difficult to present one's contemporaries with spiritual gifts. [1]

[1]See Introduction, p. 25. (Trans.)

Chinese Painting

A second great donor in Germany has fared no better—Hugo von Tschudi.[2] This man of genius presented Berlin with a great cultural treasure of paintings. The result: he was simply driven from the city. No one wanted his acquisitions. He went to Munich. The same story: no one wanted his gifts. At the Alte Pinakothek people have treated the Nemes Collection no better than a fashion show. They will breathe a sigh of relief when this dangerous collection is gone without their having been forced to keep some of it. The acquisition of a Rubens or a Raphael would be something else, indeed, for it could certainly be considered a contribution to the nation's *material* wealth.

This melancholy reflection belongs in the columns of the *Blaue Reiter* because it is symptomatic of a larger evil of which the *Blaue*

[2]See Introduction, p. 12. (Trans.)

Bavarian Mirror Painting

Pablo Picasso: *Woman with Mandolin at the Piano*

Two Drawings by Children

*Reiter* will probably die: the general indifference of people to new spiritual treasures.

We see this danger before us very clearly. Our gifts will be rejected with anger and abuse: "Why new paintings and new ideas? What can we buy with them? We already have too many old ones that we don't enjoy but that were foisted upon us by education and fashion."

But maybe we will be right in the end. People will not *want* that, but they will *have to*. For we know that our world of ideas is not a house of cards [2] to be played in, but that it contains elements of a movement whose vibrations can be felt today throughout the world.

We like to emphasize the El Greco case, because the glorification of this great master is very closely connected with the flourishing of our new ideas in art. Cézanne and El Greco are spiritual brothers despite the centuries that separate them. Meier-Graefe and Tschudi triumphantly brought the old mystic El Greco to "Father Cézanne." Today the works of both mark the beginning of a new epoch in painting. In their views of life both felt the *mystical inner construction*, which is the great problem of our generation.

59

The painting by Picasso, reproduced here (p. 58), belongs to this sequence of ideas, as do most of our illustrations.

New ideas are hard to understand only because they are unfamiliar. How often must this sentence be repeated before even one in a hundred will draw the most obvious conclusions from it?

But we will not tire of repeating it, and we will tire even less of expressing the new ideas and of showing the new paintings until the day arrives when our ideas are generally accepted.

These lines were already written when the sad news of Tschudi's death reached us.

We therefore choose to dedicate this first book to Tschudi's noble memory, as even a few days before his death he still promised us his always active support.

With burning souls we hope to continue this gigantic task, orphaned without him, of leading his people to the sources of art, in spite of our own weak resources, [3] until there shall come again a man endowed with mystical powers like Tschudi, who will honor this work and silence the impudent, vociferous opponents of that great dead man—those who negate the free spirit and deeds of excellence!

No one has experienced more painfully than Tschudi did, even after his death, how difficult it is to present one's own people with spiritual gifts, but it will be even harder for them to exorcise the ghosts that he invoked.

The spirit breaks down fortresses. [4]

A. Macke

# The "Savages" of Germany
*by Franz Marc*

In this time of the great struggle for a new art we fight like disorganized "savages" against an old, established power. The battle seems to be unequal, but spiritual matters are never decided by numbers, only by the power of ideas.

The dreaded weapons of the "savages" are their *new ideas*. New ideas kill better than steel and destroy what was thought to be indestructible.

Who are these "savages" in Germany?

For the most part they are both well known and widely disparaged: the Brücke in Dresden, the Neue Sezession in Berlin, and the Neue Vereinigung in Munich.

The oldest of the three, the Brücke, was inaugurated with great seriousness, but Dresden proved too infertile a soil for its ideas. The

E. Kirchner

time was probably not yet ripe in Germany for a more significant effect. A few years had to pass before the exhibitions of the two other groups brought new, dangerous life to the country.

Originally, the Neue Sezession was recruited in part from members of the Brücke. It was actually formed, however, by unsatisfied members of the old Secession, which moved too slowly for them; courageously the New Secessionists jumped over [5] the dark wall behind which the old Secessionists had hidden themselves, and suddenly they stood, dazzled, before the immense freedom of art. They have no program and no restraint; they want only to proceed at any price, like a river that carries along everything, possible and impossible, trusting in its own purifying force.

Lack of historical perspective prevents us from attempting here to distinguish between the noble and the weak. Any criticism we might make would concern only trifles and would stand disarmed and ashamed before the defiant freedom of this movement, which we in Munich greet with a thousand cheers.

The genesis of the Neue Vereinigung is more obscure and complex.

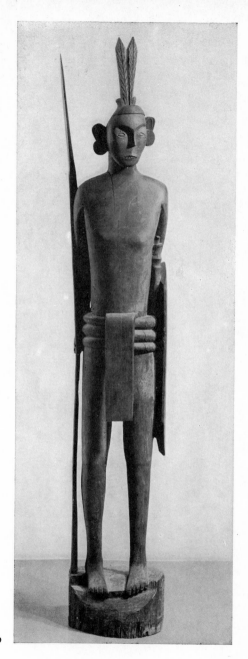

South Borneo

In Munich the first and only serious representatives of the new ideas were two Russians who had lived there for many years and had worked quietly until some Germans joined them. Along with the founding of the association began those beautiful, strange exhibitions that drove critics to despair.

Characteristic of the artists in the association was their strong emphasis on the *program*. Artists learned from each other and competed among themselves as to who understood the new ideas best. Perhaps one heard the word "synthesis" too often.

Later the young Frenchmen and Russians who exhibited with them as guests proved a liberating influence. They stimulated thought, and people came to understand that art was concerned with the most profound [6] matters, that renewal must not be merely formal but in fact a rebirth of thinking.

*Mysticism* was awakened in their souls and with it the most ancient elements of art.

It is impossible to explain the recent works of these "savages" as a formal development and new interpretation of impressionism (as B. W. Niemeyer tried to do in the statement of the Düsseldorf Sonderbund [Special Society]). The most beautiful prismatic colors and the celebrated cubism are now meaningless goals for these "savages."

Their thinking has a different aim: To create out of their work *symbols* for their own time, symbols that belong on the altars of a future spiritual religion, symbols behind which the technical heritage cannot be seen.

Scorn and stupidity will be like roses in their path.

Not all the official "savages" in or out of Germany dream of this kind of art and of these high aims.

All the worse for them. After easy successes they will perish from their own superficiality despite all their programs, cubist and otherwise.

But we believe—at least we hope we are justified in believing—that apart from all these "savage" groups in the forefront there are many quiet powers in Germany struggling with the same high, distant goals and that ideas are silently maturing unknown to the heralds of the battle.

In the dark, without knowing them, we give them our hand. [7]

# Two Pictures
## *by Franz Marc*

Knowledge must be vindicated by its children. If we wish to be wise enough to teach our contemporaries, we must justify our knowledge through our works and we must exhibit them as a matter of course.

We will make this as difficult as possible for ourselves, never fearing the ordeal by fire that will result from placing our works, which point to the future and are still unproved, beside the works of older, proved cultures. We believe that nothing can illustrate our ideas better than such comparisons. Genuine art can always be compared with genuine art, however different the expression may be. Time favors such considerations, for we believe that we stand today at the turning point of two long epochs. The awareness of this turning point is not new; its summons was even louder a hundred years ago. At that time people thought they were very close to a new era, much closer than we believe today. A century intervened, during which a long and exceedingly rapid development took place. Mankind practically raced through the last stage of a millennium that had begun [8] after the fall of the great classical world. At that time the "primitives" broke ground for a long development of a new art, and the first martyrs died for the new Christian ideal.

Today this long development in art and religion is over, but the vast land is still full of ruins, of old ideas and forms that will not give way, although they belong to the past. The old ideas and creations live on falsely, and we stand helplessly before the Herculean task of banishing them and paving the way for what is new and already standing by.

Science works negatively, *au détriment de la religion*—what a terrible confession for the spiritual work of our time.

*Reinhald das Wunderkind*

It can be sensed that there is a new religion arising in the country, still without a prophet, recognized by no one.

Religions die slowly.

But the artistic style that was the inalienable possession of an earlier era collapsed catastrophically in the middle of the nineteenth century. There has been no style since. It is perishing all over the world as if seized by an epidemic. Since then, serious art has been the work of individual artists [9] whose art has had nothing to do with "style" because they were not in the least connected with the style or the needs of the masses.[1] Their works arose rather in defiance of their times.

[1]In France, for example, Cézanne and Gauguin to Picasso; in Germany Marées and Hodler to Kandinsky. This is not supposed to be an evaluation of these artists, but is merely to indicate the development of forms of expression in France and Germany.

Kandinsky

They are characteristic, fiery signs of a new era that increase daily everywhere. This book will be their focus until dawn comes and with its natural light removes from these works the spectral appearance they now have. What appears spectral today will be natural tomorrow.

Where are such signs and works? How do we recognize the genuine ones?

Like everything genuine, its inner life guarantees its truth. All works of art created by truthful minds without regard for the work's conventional exterior remain genuine for all times.

At the beginning of this article we showed two small examples of this: a popular illustration from Grimm's *Fairy Tales* of 1832 and a painting by Kandinsky of 1910. The first is as genuine and as completely heartfelt as a folk song and was understood in its time with perfect obviousness and love; for in 1832 every journeyman and every prince shared the artistic feeling [10] out of which the little picture was created. Everything genuine created at that time had this pure, untroubled relation to the public.

But now we believe that anyone who feels the spiritual and artistic quality in the old fairy-tale picture will find in Kandinsky's painting, our modern example, the same spiritual artistic expression—even if

Campendonk

he cannot enjoy it as readily as the Biedermeier enjoyed his fairy-tale picture. For such a relationship to exist, the necessary, basic condition, even today, is that the artist's "homeland" possess a style.

Since this is not the case, a chasm *must* exist between the genuine creation of art and the public.

It cannot be otherwise because the artist can no longer create out of the now-lost artistic instinct of his people.

But could not this very fact encourage serious thinking along the lines previously suggested? Perhaps the viewer will begin to dream in front of the new painting and encourage his soul to move onto a new plane?

*The present isolation of the rare, genuine artist is absolutely unavoidable for the moment.*

This assertion is clear, only the reasons for it are missing.

The reasons, we think, are these: nothing occurs accidentally and without organic reason—not even the loss of artistic style in the

Bavarian Mirror Painting

nineteenth century. [11] This fact leads us to the idea that we are standing today at the turning point of two long epochs, similar to the state of the world fifteen hundred years ago, when there was also a transitional period without art and religion—a period in which great and traditional ideas died and new and unexpected ones took their place. Nature would not wantonly destroy the religion and art of the people without a great purpose. We are also convinced that we can already proclaim the first signs of the time.

The first works of a new era are tremendously difficult to define. Who can see clearly what their aim is and what is to come? But just the fact that they *do exist* and appear in many places today, sometimes independently of each other, and that they possess inner truth, makes us certain that they are the first signs of the coming new epoch— they are the signal fires for the pathfinders.

The hour is unique. Is it too daring to call attention to the small, unique signs of the time? [12]

Der Winter.

Bavarian Glass Painting

Mosaic in Cathedral of San Marco (Venice)

# The "Savages" of Russia

*by D. Burliuk*

Japanese

Realism changes itself into impressionism. Remaining completely realistic in art is unthinkable. In art everything is more or less realistic. But it is impossible to found a school on the "more-or-less" principles. "More-or-less" is not aesthetics. Realism is nothing but a species of impressionism. But impressionism, i.e., life seen through the prism of an experience, is a creative form of life. My experience is a transformation of the world. Becoming absorbed in an experience leads me to creative activity. Creating is at one and the same time creating experiences and creating creations. The laws of creation are the only aesthetics of impressionism and at the same time the aesthetics of symbolism. "Impressionism is a superficial form of symbolism" (Andrei Bely).[1]

"I have no doubt that from a study of the works of Raphael or Titian a more complete set of rules can be drawn than from the works

[1]One of the best known modern "young" poets (K. L.). Andrei Bely (1880–1934) was the pseudonym of Boris Nikolaevich Bugaev, b. Moscow, who studied sciences at the University of Moscow, later became an adherent of Rudolf Steiner's anthroposophic ideas, and spent many years abroad before returning to Moscow. An outstanding representative of Russian symbolism, critic, poet, and novelist, who applied musical principles to fiction. (Trans.)

72

Russian Folk Art

of Manet or Renoir, but the rules followed by Manet and Renoir were suited to their artistic temperaments and I happen to prefer the smallest of their paintings to all the work of those who have merely imitated the *Venus of Urbino* or the *Madonna of the Goldfinch*. Such painters are of no value to anyone because, whether we want to or not, we belong to our time and we share in its opinions, preferences, and delusions" (Henri Matisse).

"A renaissance is caused not primarily by perfect works but by the power and uniformity of the ideal in a generation full of life" (Maurice Denis). [13]

The congress of Russian artists, planned for December, must try above all to create an atmosphere necessary for such a uniformity. This objective, if it can be accomplished, will also unite the young artists who are not self-satisfied, but who search for new ways in art and prefer the ideal aims of international art to national and pecuniary interests.

Art is something special. If a congress were to meet for the benefit of some technical interest—aviation, navigation, auto racing, etc.—all its members would certainly admit unanimously that "we are lagging

73

V. Burliuk

behind other nations," that "compared with Western Europe, Russia lags far behind." And it would still be stated today, just as it was during the time of Peter the Great, that Western European culture should be the desirable goal for us.

Things are different in each of the spiritual disciplines and thus in painting as well. In the latter, the visible evidence of a flying airplane is lacking. Art, after all, is no Krupp cannon, which has a great deal of argumentative force. Every theoretical conceit [14] is silenced here. And, unfortunately, conceit is a characteristic Russian quality—the less culture, the greater this delusion. This delusion is naturally very comfortable: it eliminates the restless search, the restless creating, which are the greatest enemies of "Oblomovism." The art critic Alexander Benois[2] has already observed correctly that "Russian artists distinguish themselves by a dreadful laziness—yes! Russian artists are suffering from Oblomovism—and in this case they are truly national!"

Aside from this criticism, other sad aspects of contemporary Russian painting are to be observed. The earlier leaders of the *World of Art*[3]

[2]One of the important Russian Secessionists. (K. L.)
[3]The heyday of the second Russian Secession was in the 1880s and 1890s. The young radical artists of this period, who were called "decadents," grouped themselves around the magazine *World of Art*. Artists of this generation who became

gradually reached [15] the deathlike silence of the Federation, which finally sank to the level of the Wanderers. (It is known that the term "Wanderer" is used today as an invective.) In the 1890s Repin even sneered at Puvis de Chavannes and Degas, whom we find overly saccharine today. At this point the *World of Art* was still completely liberal, eagerly reproducing the French impressionists, whom I would call intimists. They are representatives of a sweet art without principles, of an art which lost ground and advanced only as far as the idea of exterior beauty and harmony of color spots. This enthusiasm for French art, however, suddenly came to an end in Russia, after which there developed a movement that paralleled French painting. In the more delicate, pure, talented minds there arose a divine light and a more conscious relationship to art. Around this light an incredible dispute developed, a veritable Walpurgis Night! The academicians were joined in this dispute by groups that earlier had been, at least outwardly, in opposition to the academy.[4] This dispute produced enough noise to drown out many troublesome questions (being asked by the thin-skinned): "Am I right after all?

better known in Germany were Somov and Serov. In 1906 Diaghilev, previously editor of the *World of Art*, organized a great Russian exhibition at the Salon d'Automne in Paris, with the above-mentioned artists the main representatives. On its way back this exhibition stopped in Berlin, where Somov left the greatest impression. When the magazine came to an end, the Federation of Russian Artists was formed, which was very much like the Federation of German Artists. The first Russian Secession started in the 1870s. It was the heyday of Russian realism. This great society staged a big touring exhibition every year, after which these artists were called "the Wanderers." One of its main representatives was Ilya Repin. (K. L.)
[4]The academic principles: "values," coloring, belief in the "realistic," "right" drawing, in a "harmonious" tone (these parts of the law are rejected by some, who, however, consider the rest to be holy all the same), construction, proportion, symmetry, perspective, anatomy (the rejection of these principles is most important, primary, most characteristic—not without good reason have even Cézanne and Van Gogh, if remotely, pointed to the necessity of liberation from this slavery!).

Russian Folk Print

Russian Folk Print

Is Apollo to be worshiped as I do? Is it really right to paint the same pictures year after year, only changing their names?" Now the game is out in the open . . .

The thing is so widespread, and Russian art is so far behind that Muther, for one, ignored it; but Benois "made up" for this omission.[5] Even Maurice Denis, despite his tact, despite his more than modest encouragement, smiled rather coolly when he was shown Russian works of art. [16]

For the followers of academic "art" the free search for beauty is nothing but "making grimaces." A patriotic success of their "genuine" Russian art would offer them the best opportunity to put their untalented "works" on the market. They are a veritable nightmare of art, the death of art. Some of them openly bare their teeth and wear their hides with dignity. They are not the most dangerous ones. The truly evil ones are the others—the wolves in sheep's clothing. O these false sheep! They are the real danger, which means—watch out!

The academicians are the real enemies of the new art, which, fortunately, does exist in Russia and which has different basic principles.

[5]A special volume by Benois which deals with Russian art, and in which the *World of Art* group occupies the leading place, was added to the Russian edition of Richard Muther's history of nineteenth-century art (first published in Munich). (K. L.)

D. Burliuk

Their representatives, Larionov, P. Kuznetsov, Saryan, Denisov, Konchalovsky, Mashkov, Goncharova, von Wisen, V. and D. Burliuk, Knabe, Yakulov, and, living abroad, Sherebtsova (Paris), Kandinsky, Werefkin, Jawlensky (Munich), revealed in their works *new principles of beauty*, as did the great French masters (such as Cézanne, Van Gogh, Picasso, Derain, Le Fauconnier, and to some extent Matisse and Rousseau).

Let the enemies of this art be convulsed with laughter. The disguised sheep shall favor us as they willingly favor a member of the *World of Art.*

There is nothing else they can do!

In order to understand the works of the artists mentioned, you have to throw the academic stuff completely overboard. Feeling must be purged, which is not so easy for those who are crammed with all sorts of "knowledge."

It is always the same old story. Even the greatest draftsmen of the nineteenth century—Cézanne, Van Gogh—had to listen to the refrain.

Our Secessionist painters are convinced to this day that Cézanne was not a bad artist, but mainly lacked the ability to draw. [17]

The newly discovered law of all the artists just mentioned is nothing but an upstanding tradition whose origin we find in the works of "barbaric" art: the Egyptians, Assyrians, Scythians, etc. This rediscovered tradition is the sword that smashed the chains of conventional academicism and freed art, so that in color and design (form) it could move from the darkness of slavery toward the path of bright springtime and freedom.

What we first thought to be the "clumsiness" of Cézanne and the frantic "handwriting" of Van Gogh is something greater after all: it is the revelation of new truths and new means.

And these truths and means are:

1. The relation of painting to its graphic elements, the relation of the object to the elements of plane (which we see signs of already in Egyptian "profile painting").
2. The law of displaced construction—the new world of construction drawing! Connected with it:
3. The law of free drawing (main representative: Kandinsky; also to be found in the best works of Denisov and especially clear in Larionov's *Soldiers*). [18]
4. The application of several viewpoints (which in architecture has long been known as a mechanical law), the combination of perspective presentation with the basal planes, that is, the use of more planes (Yakulov's *Café Chantant*).
5. The treatment of the plane and its intersections (Picasso, Braque; in Russia, V. Burliuk).

6. The equilibrium of perspectives, which replaces mechanical composition.
7. The law of color dissonance (Mashkov, Konchalovsky).

These principles are inexhaustible sources of eternal beauty. Much may be obtained here by those who have eyes to see the hidden meaning of lines, of colors. It beckons, allures, and attracts man!

Thus the chain that, because of various rules, had fettered art to the academy was shaken off: construction, symmetry (anatomy) of proportions, perspective, etc.—laws that are eventually mastered easily by the untalented—the pictorial kitchen of art!

All our expert as well as amateur critics should be the first to understand that it is high time to pull up the curtain and to open the window of true art! [19]

Most writing on art is by
people who are not artists:
thus all the misconceptions.

DELACROIX [20]

Portrait of a Stonemason (Thirteenth Century)

# Masks
*by August Macke*

A sunny day, a cloudy day, a Persian spear, a holy vessel, a pagan idol and a wreath of everlasting flowers, a Gothic cathedral and a Chinese junk, the bow of a pirate ship, the word "pirate" and the word "holy," darkness, night, spring, the cymbals and their sound and the firing of armored vessels, the Egyptian Sphinx and the beauty spot stuck on the pretty cheek of a Parisian cocotte.

The lamplight in Ibsen and Maeterlinck, the paintings of village streets and ruins, the mystery plays in the Middle Ages and children frightening each other, a landscape by Van Gogh and a still life by Cézanne, the whirring of propellers and the neighing of horses, the cheers of a cavalry attack and the war paint of Indians, the cello and the bell, the shrill whistle of the steam engine, and the cathedral-like quality of a beech forest, masks and stages of the Japanese and the Greeks, and the mysterious, hollow drumming of the Indian fakir.

Brazilian

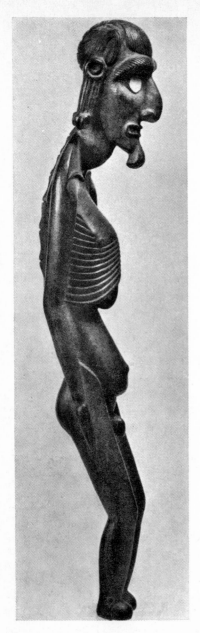

Easter Islands

Is life not more precious than food and the body not more precious than clothing?

Incomprehensible ideas express themselves in comprehensible forms. Comprehensible through our senses as star, thunder, flower, as form.

Form is a mystery to us for it is the expression of mysterious powers. Only through it do we sense the secret powers, the "invisible God."

The senses are our bridge between the incomprehensible and the comprehensible. [21]

To behold plants and animals is: to perceive their secret.

To hear the thunder is: to perceive its secret. To understand the language of forms means: to be closer to the secret, to live.

To create forms means: to live. Are not children more creative in drawing directly from the secret of their sensations than the imitator of Greek forms? Are not savages artists who have forms of their own powerful as the form of thunder?

Thunder, flower, any force expresses itself as form. So does man. He, too, is driven by something to find words for conceptions, to find clearness in obscurity, consciousness in the unconscious. This is his life, his creation.

As man changes, so do his forms change.

The relations that numerous forms bear to one another enable us to recognize the individual form. Blue first becomes visible against red, the greatness of the tree against the smallness of the butterfly, the youth of the child against the age of the old man. One and two make three. The formless, the infinite, the zero remain incomprehensible. God remains incomprehensible.

Man expresses his life in forms. Each form of art is an expression of his inner life. The exterior of the form of art is its interior.

Each genuine form of art emerges from a living correlation of man to the real substance of the forms of nature, the forms of art. The scent of a flower, the joyful leaping of a dog, a dancer, the donning of jewelry, a temple, a painting, a style, the life of a nation, of an era.

The flower opens at sunrise. Seeing his prey, the panther crouches, and as a result of seeing it, his strength grows. And the tension of his strength shows in the length of his leap. The form of art, its style, is a result of tension. [22]

Styles, also, may perish from inbreeding. The crossbreeding of two styles results in a third, a new style. The renaissance of antiquity and of Dürer, the disciple of Schongauer and Mantegna. Europe and the Orient.

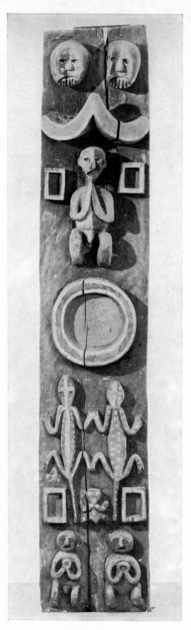

Cameroons

Mexico                    New Caledonia

In our time the impressionists found a direct connection with natural phenomena. Their rallying cry was to depict nature's organic form bathed in light, enveloped in atmosphere. It changed under their hands.

Peasant, Italian primitive, Dutch, Japanese, and Tahitian art forms became as stimulating as nature's own forms. Renoir, Signac, Toulouse-Lautrec, Beardsley, Cézanne, Van Gogh, Gauguin. They are all no more naturalists than El Greco or Giotto. Their works are the expressions of their inner lives; they are the forms of these artists' interior world in the medium of painting. This does not necessarily indicate that there is a culture, a culture that would mean to us what the Gothic style meant to the Middle Ages, a culture in which everything has form, form born from our lives—only from our lives. Self-evident and strong as the scent of a flower. [23]

In our complicated and confused era we have forms that absolutely enthrall everyone in exactly the same way as the fire dance enthralls the African or the mysterious drumming of the fakirs enthralls the Indian. As a soldier, the independent scholar stands beside the farmer's son. They both march in review similarly through the ranks, whether they like it or not. At the movies the professor marvels alongside the

Alaska

servant girl. In the vaudeville theater the butterfly-colored dancer enchants the most amorous couples as intensely as the solemn sound of the organ in a Gothic cathedral seizes both believer and unbeliever.

Forms are powerful expressions of powerful life. Differences in expression come from the material, word, color, sound, stone, wood, metal. One need not understand each form. One also need not read each language.

The contemptuous gesture with which connoisseurs and artists have to this day banished all artistic forms of primitive cultures to the fields of ethnology or applied art is amazing at the very least.

What we hang on the wall as a painting is basically similar to the carved and painted pillars in an African hut. The African considers his idol the comprehensible form for an incomprehensible idea, the personification of an abstract concept. For us the painting is the comprehensible form for the obscure, incomprehensible conception of a deceased person, of an animal, of a plant, of the whole magic of nature, of the rhythmical.

Does Van Gogh's portrait of Dr. Gachet not originate from a spiritual life similar to the amazed grimace of a Japanese juggler cut

Child's Drawing: *Arabs*

in a wood block? The mask of the disease demon from Ceylon is the gesture of horror of a primitive race [27] by which their priests conjure sickness. The grotesque embellishments found on a mask have their analogies in Gothic monuments and in the almost unknown buildings and inscriptions in the primeval forests of Mexico. What the withered flowers are for the portrait of the European doctor, so are the withered corpses for the mask of the conjurer of disease. The cast bronzes of the Negroes from Benin in West Africa (discovered in 1889), the idols from the Easter Islands in the remotest Pacific, the cape of a chieftain from Alaska, and the wooden masks from New Caledonia speak the same powerful language as the chimeras of Notre-Dame and the tombstones in Frankfurt Cathedral.

Everywhere, forms speak in a sublime language right in the face of European aesthetics. Even in the games of children, in the hat of a cocotte, in the joy of a sunny day, invisible ideas materialize quietly.

The joys, the sorrows of man, of nations, lie behind the inscriptions, paintings, temples, cathedrals, and masks, behind the musical compositions, stage spectacles, and dances. If they are not there, if form becomes empty and groundless, then there is no art. [26]

89

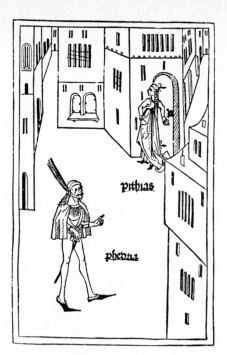

German (Fifteenth Century)

# The Relationship to the Text
## by Arnold Schönberg

Comparatively few people are able to understand what music has to say from a purely musical point of view. The following assumption is widespread as only false and trite assumptions can be: music has to convey some sort of idea; if that is missing, the piece of music has either not been understood or it is regarded as worthless. Nobody would demand anything similar from any other art, but people are content with the effects of the material. And the subject, the depicted objects in these arts [27] satisfy the limited intelligence of the intellectual bourgeoisie. Music lacks such an immediately perceptible subject, so some look for purely formal beauty behind its effects, others for poetical developments. Schopenhauer himself first expressed a wonderful insight into music: "The composer reveals the innermost essence of the world and pronounces the most profound wisdom in a language that his reason cannot understand; he is like a mesmerized

Ivory Sculpture

A. Kubin

somnambulist who reveals secrets about things that he knows nothing about when he is awake." Even Schopenhauer later gets lost when he tries to translate details of musical language, *which reason cannot understand*, into the language of our concepts. He should have realized that [28] in this translation into concepts, into the language of man, which is abstraction, reduction to the visible, the essence is lost, the language of the world, which perhaps has to remain unintelligible, only perceptible. But Schopenhauer was justified in reasoning this way. As a philosopher it was his aim to put down in concepts the essence of the world, the immeasurable wealth. The concepts, of course, soon enough proved their poverty. Wagner, too, was right when he attributed literary programs to Beethoven's symphonies so that the average person could get an indirect idea of what he (Wagner) as a composer had envisioned directly.

This procedure becomes fatal when it is applied too generally. Its value is turned into its opposite: people look for actions and feeling in music as if they had to be there. In Wagner it is, in reality, different: the impression of the "essence of the world" that he conceived in music becomes productive in him and stimulates a re-creation in the material of another art. But the actions and feelings that occur in this poetry were not in the music; they are just the building materials that the poet used, because poetic art, being bound to matter, does not possess such a direct, unclouded expression.

Bavarian
Glass Painting

This capacity for pure vision is very rare and only to be found in highly cultured people. It is understandable that the most uninformed in art are lost when a few accidental difficulties bar their way [29] to the enjoyment of music. Our scores are more and more difficult to read; they are given comparatively few performances and these often are so fast that even the most sensitive listener receives nothing but a fleeting impression. This makes it impossible for the critic to fulfill his task honestly. He has to report and to judge, but he is frequently unable to form any notion of the score. He might well decide to be honest, if it would do him no harm. He helplessly faces the pure musical effect and he prefers, therefore, to write about the kind of music that is somehow related to a text, about program music, lieder, operas, etc. This could almost be excused because opera conductors

93

themselves, when asked about the music of a new opera, chatter almost entirely about the libretto, theatrical effects, and actors. Since musicians and composers have become educated and think they have to prove this by shying away from talking shop, there are hardly any musicians left with whom you can talk about music! But Wagner, whom the musicians like to refer to, wrote a great deal on purely musical subjects. I am sure he would absolutely disagree with these results of his misunderstood endeavors.

It is therefore nothing but an easy way out of this dilemma when a critic of music writes about a composer that his composition did not do justice to the words of the poet. The "amount of space" in the magazine is always too limited for the inclusion of necessary proofs; this, of course, makes up for the critic's lack of ideas, but the artist is the one found guilty because of "insufficient evidence." When the evidence for such statements is submitted [30], it is a witness to the opposite. The critic only says how someone who cannot write music would do it, and what music would be like if it were written by an artist. This is even true when a composer writes criticism. Even if he is a good one. For at the moment that he is writing criticism, he is not a composer: he is not *musically inspired*. If he were inspired, he would not describe how the work should be composed; he would compose it. This is even faster and easier and more convincing for the one who knows how to do it.

Such judgments actually originate in a very trite idea. In a traditional scheme a certain incident in the poetry must be absolutely parallel with a certain loudness or speed in music. *This* parallel may also occur, may even be much *more profound*, when externally the opposite is true—when a delicate idea is rendered by a fast and vigorous theme. A subsequent vigor can develop from it more organically; but aside from that, such a scheme is surely contemptible because it is traditional. Moreover it makes us think of music as a language that is "creating and thinking" for everybody. And the application of these ideas by critics leads to such results as an essay I once read somewhere: "Wagner's Mistakes in Declamation," in which a fathead showed how *he* would have composed certain passages if Wagner had not jumped the gun on him.

A few years ago I was deeply ashamed to discover that in some of Schubert's lieder, which I knew well, I had no idea what was actually happening in the poems they were based on. But when I read the poems I found that I had gained nothing in the understanding of the

Egyptian

lieder, that they did not in the least influence my opinion of the musical statement. On the contrary, I had quite obviously grasped the content, the real content, [31] perhaps even better than if I had clung to the surface of the actual verbal ideas. Even more decisive than this experience was the fact that, intoxicated by the sound of the first words in the text, I had finished many a lied of my own without in the least caring for the further development of the poem, without even noticing it in the ecstasy of composing. Only some days later did I think of looking up the poetical content of my lied. To my great amazement I realized that I could not have done more justice to the poet. The direct contact with the sound of the first words made me sense what necessarily had to follow.

From this I concluded that a work of art is the same as any perfect organism. It is so homogeneous in its composition that it reveals its true inner essence in each detail. If you cut any part of the human body, the same blood will flow. If you hear one verse of a poem, one measure of a piece of music, you are able to comprehend the whole. In the same way, one word, one glance, one gesture, the gait, even the color of the skin is enough to distinguish the character of a man. [32] I had completely understood Schubert's lieder—including the lyrics —through the music alone, Stefan George's poetry through the sound alone. My understanding was so complete that it could hardly have been equaled and never have been surpassed by analysis and synthesis. Later on, such impressions frequently appeal to the intellect. The intellect is asked to make these impressions suitable for daily use, to dissect and to classify, to measure and to examine, to dissolve the

Marquesas Islands

Bavarian Glass Painting

R. Delaunay: *The Eiffel Tower*

El Greco: *St. John*

*Horses* in water color by F. Marc [only in the de luxe edition and second edition]

*Horses* in water color by F. Marc [In original edition only; changed in second edition]

unusable whole into details that can be expressed at any time. However, even artistic creation often takes this detour before it reaches its proper conception. But there are signs that even the other arts, which apparently are closer to the subject matter, have overcome the belief in the power of intellect and consciousness. Karl Kraus calls language the mother of ideas; Kandinsky and Oskar Kokoschka paint pictures in which the external object is hardly more to them than a stimulus to improvise in color and form and to express themselves as only the composer expressed himself previously. These are symptoms of a gradually spreading recognition of the true essence of art. And with great joy I read Kandinsky's book, *Concerning the Spiritual in Art*, in which a way is shown for painting that arouses hope that those who demand a text will soon stop demanding.

It will also make clear what was already clear in another context. Nobody doubts that a poet working on a historical scene may move with the greatest freedom; nobody doubts that it is unnecessary for a contemporary painter, who wants to paint historical events, to compete with a professor of history. We want to see what the work of art has to give and not its external stimulus. With compositions based on poetry the exactness of rendering the action is as irrelevant to its artistic value as the resemblance to the model is for a portrait. A hundred years later no one will be able to check the likeness, but the artistic effect will always remain. This effect will exist not because a real man, the man apparently depicted, appeals to us—as perhaps the impressionists believe—but because the artist appeals to us, the person who expressed himself here and who must be recognized in the portrait in a higher degree of reality. Once this has been recognized, it is easy to understand that the external congruence of music and text, which reveals itself in declamation, tempo, loudness, has as little to do with the internal congruence, and stands at the same level of primitive imitation of nature, as the copying of a model. An apparent divergence on the surface can be necessary because of a parallel movement on a higher level. Therefore, the evaluation of music according to the text is as reliable as the evaluation of albumen according to the characteristics of carbon. [33]

Wooden Figures (Malayan)

# "The Wreath of Spring"[1]
*by M. Kuzmin*

Birth, the last farewell—everything the Lord of Heaven gives.
Voice of the frog, jasmine blossom—everything the Lord of Heaven
  gives.
Heat of summer, flowers of spring, grapes of bronzed fall,
Avalanches in the mountains—everything the Lord of Heaven gives.
Profit and ruin, on the journey happiness and death,
Royal power and cobwebs—everything the Lord of Heaven gives.
To the cupbearer the smile in the face, to white hair the honor of a sage.
Erect stature and hunchback—everything the Lord of Heaven gives.
River Euphrates, prison towers, walls of rock, expanse of deserts,
Everything my eye can see—everything the Lord of Heaven gives.
It is my fate—power of laughter, sounds of meeting and departure,
I do not curse my fate—everything the Lord of Heaven gives. [34]

[1] English translation from Kandinsky's German version. (Trans.)

# Signs of Renewal in Painting

*by Roger Allard*[1]

Cézanne

[1]English translation from
the German version of
the original French. (Trans.)

The best training for a study of the pattern of artistic evolutions is to experience the development of a style in painting up to its fossilization and death, preferably a style that is survived by the pseudostyle of its imitators. [35]

Because impressionism now belongs to the past, we can detect its historical relationship not only to the period directly preceding it—which frequently has been done before—but also to the period following it, the art of our time.

The undeniable analogy of impressionism to naturalism, especially in its late period, is perhaps the main reason that impressionism could not lead to a more important style. In all other periods great epochs of art founded schools and the forms of great renovators created definite styles. The narrow limits of the impressionist principle allowed only three or four great artists to develop their personalities.

Surprising reactions followed, such as neoimpressionism, which basically was nothing but a concealed restoration and which finally developed into an art for gourmets. Even the great heritage of Cézanne was dissected and the labor of his discoveries became a cheap commodity. But Cézanne's art is the arsenal out of which contemporary painting takes its swords for the first parry and thrust in the struggle to defeat naturalism, false literature, and pseudoclassicism. Today different things are being fought for.

It seems important to us to outline this short prehistory of "cubism" up to its final definition in art, because we intend to combat the trivialities and false reports that are disseminated today in all magazines and newspapers on infamous cubism.

What is cubism?[2]

First of all, it is the deliberate intention to restore again to painting the knowledge of measure, volume, and weight.

Instead of the impressionist illusion of space based on perspective in atmosphere and a naturalism of colors, cubism renders simple, abstract forms in defined relationships and proportions to one another. The first postulate of cubism is, therefore, the order of objects—not however, the order of naturalistic objects, but of abstract forms. It perceives space as a compound of lines, units of space, quadratic and cubic equations and relationships. [36]

---

[2]Cf. several articles on that subject in *L'Art Libre* (November 1910), *Les Marches du Sud-Ouest* (June 1911), *La Revue Indépendante* (August 1911), *La Côte* (October–November 1911), etc.

Le Fauconnier

The task of the artist is to create an artistic order out of this mathematical chaos. He wants to awaken the latent rhythm of this chaos.

Any view of life in which the most dissimilar forces struggle in opposition is a center for this idea. The external subject of this view of life is only a pretext for, or, rather, the argument of the equation. This has never been different in art—only this final meaning has been hidden for centuries; modern art is trying to rediscover it.

Isn't it strange how difficult it is for our contemporary critics and aesthetes to admit the necessity for a re-evaluation of nature in terms of an exact and abstract world of forms in the visual arts? In other fields, in [37] music and poetry, a similar abstraction is a self-evident postulate for them. Camille Mauclair, for example, regards cubism as a scholastic sophistry leading to the sterilization of the creative urge. He forgets that two creative people are necessary to enjoy a work of art: the first is the artist who creates it, the great stimulator and inventor, and the second is the viewer whose mind has to return to nature. The further the two of them advance to reach the same goal, the more creative they both are.

Everyone who is fully aware of his own aesthetic feelings will sense the truth of this statement.

If the possibility of a legitimate explanation of aesthetic actions is denied right at the beginning, then one can well ask whether order

Henri Matisse: *La Danse*

equals disorder. Cannot or must not the human mind enter here with theories and differentiations? This is not at all to deny that all values in this world, especially aesthetic values, are relative and changing.

It is hardly necessary to mention the forms and means of expression that have been applied so far in the wide system of cubism; the development of these ideas is still at its beginning.

Some dematerialized their conception of the world in the attempt to separate its various parts from one another. Others searched for a system to transform objects until they became abstract cubist forms and formulas, weights, and measures.

There was the idea of the futurists ("*le cinématisme*") who wanted to replace the old European rules of perspective by a new, to some extent centrifugal, perspective that would no longer show the viewer one fixed plane, but would, so to speak, lead him around the object. Others tried a reciprocal interpenetration of objects to increase the expression of motion.

None of these ideas, however, alone or combined, produces an artistic canon for creating. The terrifying fertility of these ideas is rather the evil sign of decadence. No sound artistic structure can be erected on such a basis. In contemporary literature we have the same movement: the ardent urge for synthesis drives the spirit to arbitrary expositions that push poetry to the picturesque, which is, at the same time, on the brink of the carefully avoided anecdote; a dangerous vicious circle [38] that banishes the spirit. Therefore futurism appears to us as a malignant growth on the healthy tree of art.

We are proclaiming the right of the new constructive movement in art. We defend its good purpose against all romantic points of view which deny the sleeper the opportunity to think and to speculate and in which he is regarded as an inspired dreamer whose left hand does not know what his right hand is doing.

The first and most noble right of the artist is to be a conscious builder of his ideas.

Many highly talented artists of our time bleed to death disregarding their right to work *consciously*. The most honest among them realizes the imperative necessity of new aesthetic laws and the knowledge of these laws.

Cubism is no new fantasy of "savages," no scalp dance around the altars of the "officials," but an honest search for a new discipline.

From this point of view the old quarrel about the priority of its discovery is completely irrelevant. Unquestionably Derain, Braque, and Picasso made the first formal attempts in cubism; these were followed by more systematic works by others. We are indebted to the spirit and temperament of these artists; therefore we place their names on the top of this objective essay, which is not meant to evaluate talents but only to show the line of development. Since the exhibition in the Salon d'Automne in 1910, this line took a very clearly defined direction. The exhibition of the Indépendants and of the Salon d'Automne in 1911 finally made it clear that one could speak of a "renewal" in painting. The critics' attitude was remarkable; they were no longer reserved but very angry and offensive. They would have put up with the works of some amateurs, but they could not bear to be confronted with a serious movement that threatened to cause the final downfall of their own senile art. They clearly expressed their indignation that these new painters did not even think of wiping the sand out of their tired eyes so that they could see the athletic qualities of these new paintings. The critics were upset about the vast, pulsating life of this rich movement; they would rather have tried to classify it as a hoax upon the masses. [39] This was a waste of time, for among the new painters were celebrities such as Metzinger, and Le Fauconnier, who asserted the noble reserve of his northern character in the delicate architecture of his space creations. Then Albert Gleizes, who forced his rich imagination into logical constructions. Then Fernand Léger, indefatigably searching for new relationships of proportions; finally the painterly Robert Delaunay, who conquered the arabesques of the plane and who shows the rhythm of great, infinite depths. [40]

This group pursues its high aims apart from the dreamlike art of Rouault and without connection to the talented and delicate Matisse. To this group belongs also the stylish Marie Laurencin and R. de La Fresnaye, full of tradition and schooling and full of new forms. Is it not an infallible proof of the power of the movement that from an inner congeniality it attracts the most heterogeneous elements? Dunoyer de Segonzac, L.-Albert Moreau, Marchand, Dufy, Lhote,

Embroidery (Fourteenth Century) ▶

Cézanne

Marcel Duchamp, Boussingault, and many other names whom I soon
hope to have the opportunity to speak about.

With the sculptors, like Duchamp-Villon and Alexander Archipenko,
we also experience the trend toward new ideas.

The new spiritual movement is also no longer solely French. The
same search for renewal in art resounds abroad.

Is it possible to see in this valiant movement anything else but the
rebellion against worn-out aesthetics and at the same time the creation
of a new canon, which shall give our life style and inner beauty? [41]

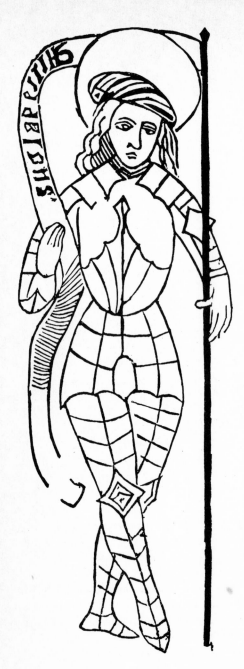

In 1807 Goethe said: " . . . in painting the knowledge of the thorough bass has been missing for a long time; a recognized theory of painting, as it exists in music, is lacking." [42]

From *Goethe im Gespräch*, p. 94, Insel-Verlag, 1907.

Wooden Figure (Malayan)

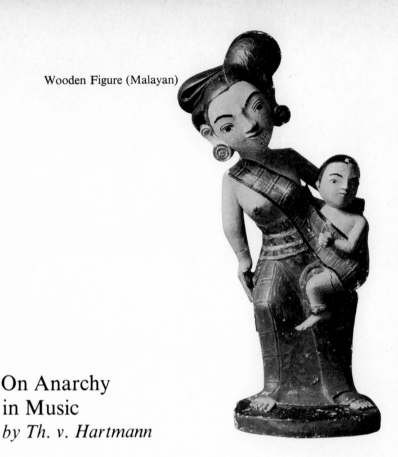

# On Anarchy
# in Music
## *by Th. v. Hartmann*

External laws do not exist. If the inner voice does not rebel, everything
is permitted. This is generally (and thus in art) the only principle of
life that was proclaimed in former times by the great adepts of the
*verbum incarnata.*

In all the arts, and especially in music, every method that arises
from an inner necessity is right. The composer wants to express what
at the moment is the intention of his intuition. At this moment he
might feel the need for a combination of sounds, which, according to
present theory, is regarded as cacophonous. It is obvious that such a
judgment of theory cannot be considered an obstacle in this case. The
artist is compelled to use such a combination because its use was
determined by his inner voice: the correspondence of the means of
expression with inner necessity is the essence of beauty in a work. The
work's persuasive power, which entirely depends on this correspond-

113

A. Macke

ence, [43] finally forces the audience to accept the beauty of the work, despite the new methods. If we make this point of view the principle of our judgments, we overcome the difficulties of artistic evaluation of contemporary anarchists in music. These are composers who know of no external limits and only listen to their inner voices when expressing their artistic egos.[1]

Any combination of sounds, any sequence of tone combinations is possible. But here we encounter the great problem that exists not only in music but in all the arts. All methods can be equally right; but will they all together have the desired effect on *the* particular sense organ appropriate to each particular art? In other words, can the laws of sense perception establish rigid limits for absolute freedom in the choice of methods? These laws are often invincible. Frequently, it can be observed that a transgression of them results in the dominance of unimportant parts, which silence the main element of a work. This is caused by a mysterious struggle [44] among the methods used, which interferes with the total effect of the methods on our sense organs. Suppose it is necessary for a composer to shock the audience

---

[1]It seems to be appropriate to mention that the use of new methods is, of course, no satisfactory measure of artistic value. On the other hand, a highly artistic work could be written today that, despite our drive for new methods, does not go beyond the borders of classical forms.

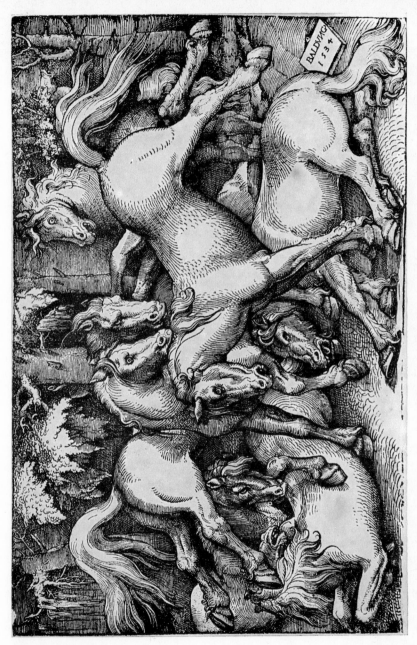

Baldung Grien: Woodcut

A. Bloch

by a certain strange combination of sound. To achieve this, it is obvious that a series of different sound combinations must necessarily have preceded this main combination; otherwise our ear would get used to the method we have selected for our purpose, and we would no longer be able to react strongly to the necessary combination.

I am expecting the reply: we cannot rely on our sense organs; they are imperfect. Moreover, from the start they are unconsciously accustomed to certain formulas that we consider to be axiomatic. On the other hand, like everything else in the world, hearing can be developed. What musicians in former times thought wrong is harmony to the contemporary ear.

To this I reply as follows: everything undoubtedly develops and is subject to certain changes. Everything aims at an ideal, an infinitely distant point. In the same way the laws of hearing will infinitely develop and improve. Undoubtedly the germ of the future ideal function of hearing is already present in the ears we now have, and the

E. Heckel

laws of an ideal hearing will basically be related to the laws of our present hearing, despite its great distance from our time. On the other hand, [45] what we think to be a law of our hearing is often not a law at all. In studying these laws, theoreticians frequently consulted not the master in this field, our ear, but its neighbor, our reason, which unfortunately is not very competent in the field of art. This often had very sad consequences that could be felt during whole epochs of musical development. Just remember the time when theoreticians considered sixths and thirds to be forbidden dissonances. Nor does our time excel in profound knowledge of the laws of hearing—this can be proved by the complete *unreliability* of present musical theory during the last ten years as far as new sounds are concerned. But this unreliability cannot undermine the belief in new explorations into the law of hearing because these laws undoubtedly exist.

Now we reach the Gordian knot. The question is: does our inner voice in its creative power infallibly correspond to the real laws of our senses (in our case—the ear), or does our unconscious creative force sometimes fail to translate our intentions clearly into the language of feeling? Is it necessary for an entirely clear translation to have the conscious assistance of the laws of the senses? To me, the engagement of the conscious element seems to be necessary, absolutely necessary, but only to enrich creative [46] methods; i.e., only if this conscious element provides new possibilities, discovers new worlds. Herein lies the great hope for future musical theory, as well as for the other arts; a theory that does not want to promulgate the tedious "one can" or "one cannot," but does say, "In this case one can use this, or that, or yet another method." These methods will perhaps be related to earlier ones, but they will possibly reveal much more efficient possibilities than those that are made available to us by the unconscious feeling only.

The principle of anarchy in art should be welcomed. Only this principle can lead us to a glorious future, to a new Renaissance. But this theory should also not turn its back on other courageous path-finders. By discovering the new laws, art should rather lead to an even greater, more conscious freedom—to different new possibilities. [47]

Russian Folk Print

Greek

# Robert Delaunay's Methods of Composition

*by E. v. Busse*

Delaunay's paintings may appear bizarre, or, at the very least, enigmatic to those who look at them unprepared. This reaction, as we shall try to prove, is not the fault of the paintings but is the result of the viewer's prejudice or his different attitude toward art. Therefore, we want not to "criticize" the paintings but to analyze the artist's intentions, the ideas he expresses in his paintings. We shall try to explain in colloquial language the process of forming ideas that the artist, by *his* own means of expression, achieved in the paintings. We want to give the kind of "interpretation" that is more familiar to the public. Delaunay himself does not give us this interpretation. He is a painter and concentrates solely on painting. Painting absorbs all his means of expression and does not leave him any other possibility. What at first glance appears to be a limitation is really with him a [48] strength, and consequently his ideas find their most natural and at the same time their most perfect realization in painting.

Delaunay has not always been an abstract artist. His first works were confined to simple reproductions of external nature. He acquired skills that ensured him against subsequent technical difficulties. His essential artistic development began only when he was able to disci-

Delaunay: *St. Séverin*

pline his natural talent and devote it to his desired expression. In his first paintings the arrangement of colors did not correspond to the treatment of lines; despite their color effects these pictures were flat and dull. This led him to the problems of perspective and space. At the same time his sensitive faculties deepened, and he recognized his artistic mission in portraying what nature yielded to his *feelings*. This he wanted to make visible in a form that other people could understand. His aim was no longer a reproduction and imitation of objective nature but the embodiment of the idea that had come to him when he contemplated nature.

This kind of creation is original and demands a new form of expression, an idea characterizing the *means*, to realize it. The search for a means of expressing one's ideas faithfully and down to the smallest detail of technique is the leitmotiv in the development of an artist. At first, form is most important. The subject plays an absolutely subordinate role, but it has a close connection to the idea that it expresses. The artist always chooses a subject that can easily render the idea.

The first stage of this development is represented by the painting *St. Séverin*. The artist intends to focus the viewer's attention on the center. He achieves this not by content or objects (objects moving to a particular point) but by an adequate dynamics of space. This is created by a proportionate distribution and correspondence of colors, as well as by curving lines corresponding to the motion. Delaunay goes back to Delacroix's painting *The Entrance of the Crusaders into Constantinople* or, to be more exact, to a much freer sketch for this painting. The dynamics of space are created here by a latent movement of masses, not by objects frozen in postures of movement. All the lines—even the streets of the town in the background—correspond to this movement. Although stressing subject matter in his painting, Delacroix was forced to make concessions to historical [49] and decorative demands, that is to say, extra-artistic demands. Delaunay remains consistent in his radical execution of the motif. There is just one inconsistency in his painting: not all the means express his idea; some of them reproduce nature in simple imitation. Besides his idea, [50] he paints the physical aspect of the cathedral, the exterior of the natural object. The viewer easily recognizes it as a nonabstract "subject," which does not serve the idea but disturbs it. It confuses him and keeps him from understanding clearly the representation of the idea. To the majority of viewers this painting will, therefore, appear only as a distorted reproduction of nature.

In order to eliminate this disturbing distraction from the main idea, the initial method of composition must be continued logically: the imitative reproduction of exterior nature must be avoided, and for it factors must be substituted that render only the latent law of nature. The articulation of this law should arouse in the viewer a feeling corresponding to the impression of nature analogous in a way to the colloquial term "borrowed concept." In his *Eiffel Tower* Delaunay tries to solve this problem. For this purpose he shatters the optical image of nature and dissects it into small pieces. Size, color, and arrangement of the pieces are again controlled by the aforementioned dynamics of space. Even at this stage of development the artist is inconsistent in that the small parts of the dissected image are still fragments of the optical image of nature, each of which represents in itself an imitation of nature. He does not yet dare to draw the final consequences. The point is to form these pieces in such a way that they are represented by different technical methods.

In the picture that represents his next stage of development, *La Ville*, this new method appears in the form of a geometric cube. All exterior objects are dissolved, i.e., transposed into this form. The problem of space dynamics is not solved here. The rhythm of the latent movement does not yet include all *directions* of movement. Also, the dominating vertical movement impairs the balance of the movements as a whole. A complete equilibrium of all factors is still missing.

The problem is solved in a second painting of *La Ville*. Retaining the technical expedient of the cube, a further development is achieved through the balance of movement in all directions. The movements downward and upward are juxtaposed with a corresponding horizontal movement, a circling one, and, finally, a concentric movement. This concentric movement gives the painting the quality of a complete self-contained whole. Without it, there would be only a fragmentary space section that could be extended at will over the frame of the picture. The concentric movement is achieved by an increasing refraction of all lines toward the center of the painting, i.e., through a reduction of the cubes. The treatment of color supports the space dynamics created by the construction of lines in that the arrangement of color tones also suggests the movements to the viewer. [51]

With the second *Ville* Delaunay's *formal* development reached a stage at which he could stop for a while. He began to make use of his experiences in other aspects of painting. First he used the "landscape," or only a certain part of it, to demonstrate his ideas; then he decided

Delaunay: *The Window on the City*

to extend the ideas to everything that the eye and mind can conceive: the inherent laws of everything that exists and its subjective understanding and representation. [52]

E. Kahler

# Eugen Kahler: An Obituary
## by Wassily Kandinsky

On December 13, 1911, Eugen Kahler, just thirty years old, died in Prague. Death took him tenderly into its arms, without suffering, without anything horrible or ugly. We might say that Kahler died biblically.

And thus Kahler's death corresponded to his life.

He was born into a wealthy Prague family on January 6, 1882. He attended a Gymnasium for five years and after that a business college, a fact that seems incredible to us today: Kahler's soul was so removed from anything practical; he lived so entirely in the land of his dreams. As early as 1902 he was to go to an art academy in Munich, but he contracted nephritis and was operated on in Berlin. Despite this first sign of his disease he devoted himself entirely to the study of art. Two years in the Knirr School, one year with Franz Stuck at the Academy in Munich, one year with Habermann, and then Kahler felt strong enough to seek his way alone.

His inner voice was so clear, distinct, and precise that he could rely on it utterly. Travel to various cities and countries (Paris, Brussels,

Berlin, London, Egypt, Tunis, Italy, Spain) was actually always the same journey in the same country. It was always in the same world which must be called Kahlerland. Now and then the nephritis affected him, and sometimes Kahler had to stay in bed for weeks. But he remained entirely the same: lying in bed, he drew and painted his dreams, he read a great deal and continued to lead his strangely intense inner life. [53] In London, for example, many typical Kahler water colors were painted, rewarding enough as the life's work of any artist. It was the same in the winter of 1911 in Munich, where, lying feverish in a sanatorium, he painted another long series of marvelous water colors. This continued: traveling from one sanatorium to another, Kahler remained true to himself to his last breath.

Kahler's delicate, dreaming, serene soul, with its pure Hebrew cast of unappeasable mystical sadness, was afraid of one thing only: the ignoble. And his entirely *noble* soul did not seem to belong to our time.

Kahler left numerous oil paintings, water colors, drawings, etchings.

About a year and a half ago he had a small exhibition in Thann-hauser's Moderne Galerie in Munich; in the usual manner the critics received it haughtily and with didactic admonitions.

A great number of profoundly felt poems, of which he had never spoken, were found after his death. [55]     K.

Kahler

Cézanne: *Still Life*

# Scriabin's "Prometheus"
## by L. Sabaneiev

In analyzing Scriabin's work it is difficult to distinguish its individual forms from the general idea, from the ultimate "artistic idea" that has now become completely explicit in the composer's consciousness. This artistic idea is a positive, mystical action that leads to an ecstatic experience—to ecstasy, to the perception of more elevated dimensions of nature. We see the logical development of this idea, beginning with Scriabin's First Symphony up to his *Prometheus*. His First Symphony is a hymn to art as religion, his third represents the liberation of his mind from its chains, his confidence in his own personality, a poem of ecstasy, of the joy of free action, the ecstasy of creating. These are different stages of development of one and the same idea, which found its complete embodiment in Scriabin's mystery, in an overwhelming ritual whose ecstatic ends are served by all methods, all "caresses of the senses" (from music to dance, from the play of light to symphonies of smells). If one penetrates deeply enough into Scriabin's mystical art, one realizes that there is neither reason nor right to classify it as music alone. Mystical-religious art, which expresses all of man's secret abilities and leads to *ecstasy*, has always used *all* available means to affect the soul. We find as much in our contemporary church service— the descendant of classic mystical ritual—on a smaller scale the idea of uniting the arts is preserved. [57] Don't we find there music (singing, sounds of bells), plastic movements (kneeling, ritual of the priest's

German

Benin

V. Burliuk: *Landscape*

actions), play of smells (incense), play of lights (candles, lights), painting? All arts are united here in one harmonious whole, to attain one goal—religious exaltation.

This exaltation is achieved despite the simplicity of the methods used. Of all the arts employed in the present-day church, only music has had a great, marked development; the others are weak, almost attenuated. The various branches of the arts have, since the time of their religious interaction in antiquity, become autonomous and have reached a startling perfection independent of each other. Music and the art of words have reached the highest point; however, of late the arts of movement and of the pure play of light—the symphony of colors—have begun to develop. Today we encounter more and more attempts to vivify the art of movement. The methods of many innovators in painting can be called only an approach in painting to the pure play of colors.

The time for the *reunification* of the separate arts has arrived. This idea was vaguely formulated by Wagner, but Scriabin expresses it much more clearly today. All the arts, each of which has achieved an enormous development individually, must be united in one work, whose *ambiance* conveys such a great exaltation that it must *absolutely* be followed by an authentic ecstasy, an authentic vision of higher realities. [58]

All the arts are not equal in this unification. The arts that will dominate are those directly based on impulses of the will, which can express the will directly (music, word, plastic movement). Arts that do not depend on impulses of the will remain subordinate (light, smell); their purpose is resonance, strengthening the impression of the primary arts. They are still undeveloped arts that obviously cannot exist independently, without the support of those that are primary.

While the idea of the "mystery," i.e., the idea in general, remains unrealized, however, a *partial* unification of the arts is advisable (even if one begins with only two arts). This is what Scriabin attempts to do in his *Prometheus*. He unites music with one of the "accompanying" arts—the play of colors, which has, as could be expected, a very subordinate position. In *Prometheus* the symphony of colors is based on the principle of corresponding sounds and colors, as we elaborated in *Music*.[1] Each key has a corresponding color, each change of harmonies has a corresponding change of colors. All this

---

[1]In this article (*Music* [Moscow], January 1911, No. 9, 199) Sabaneiev says: Scriabin's musical color sensations could represent a theory of which the composers could gradually become aware. This is the chart:

| | | | |
|---|---|---|---|
| C | Red | F-sharp | Blue, intense |
| G | Orange-pink | D-flat | Purple |
| D | Yellow | A-flat | Red-purple |
| A | Green | E-flat | Steely |
| E | Whitish-blue | B-flat | With a metallic shine |
| B | Similar to E | F | Red, dark |

When arranging these tones on the circle of fifths, the regularity becomes obvious. The colors arrange themselves almost exactly according to the spectrum. Deviations depend only on the intensity of feeling (e.g., E-major—moon-colored). E-flat and B-flat are not to be found in the spectrum; according to Scriabin these tones have an indistinct color, but a very distinct metallic shine. This correspondence of colors was used by Scriabin in his *Prometheus*. Those who listened to the *Prometheus* with the corresponding light effects admitted that the musical impression was in fact absolutely equaled by the corresponding lighting. Its power was doubled and increased to the last degree. This happened despite a very primitive lighting, which produced only an approximation of the colors! At the end Sabaneiev remarks that this problem of color-sounds "can be solved only by a detailed study, by collecting and studying the statistical material." This was done in the "no longer distant future" (p. [109]). (K. L.)

Paul Gauguin: Wood Relief

Classical Relief

is based on A. N. Scriabin's intuition of color-sounds. In his *Prometheus* the music is almost inseparable from the harmonies of color. Strange, caressing, and at the same time deeply mystical harmonies emerged from these colors. The impression produced by the music is indescribably strengthened by the play of colors. The organic depth of this "whim" of Scriabin and its aesthetic logic becomes obvious.

Let us study the musical aspect of *Prometheus*. In *Music*[2] I have already had the opportunity of pointing out that *Prometheus* represents a crystallization of Scriabin's style in his latest period. Ever since his first composition Scriabin has looked incessantly for corresponding sounds, for [60] mystical sounds that could embody his ideas. It is not hard for a specialist in his work to trace the evolution of the typical Scriabin harmonies from his first work to *Prometheus*. This evolution has followed an entirely intuitive path. Only in his latest work has he become conscious of the harmonics principles that he

[2] See No. 1 of the magazine *Music*, published in Moscow.

had earlier used unconsciously. It is impossible not to recognize here the startling characteristics of his musical intuition. Isn't it wonderful that Scriabin can sometimes unconsciously, without any "theoretical" intention, find appropriate harmonies that are at the same time subordinated to a strict regularity and can be found within the boundaries of a certain scale, of a certain musical principle? Here is this scale, which consists of six sounds, and the basic harmony, which consists of six sounds out of this scale, arranged in fourths:

In this harmony and in its arrangement a great variety of intervals is to be found: the perfect fourths, E–A, A–D; the augmented fourths, C–F-sharp, B-flat–E; and the diminished fourth, F-sharp–B-flat. The scale itself, C D E F-sharp A B-flat, is acoustically justified. These sounds are overtones of the so-called harmonic scale of sounds, that is, sounds whose harmonics are a series of progressing numbers.

1, 2, 3, 4, 5, 6, 7, 8, 9, 10, 11, 12 . . .

The afore-mentioned scale. (C D E F-sharp A B-flat) consists of the sounds 8, 9, 10, 11–13, 14. It may be concluded that theoretically in this case we do not find the real F-sharp, A, and B-flat we know, but different ones; they all sound deeper than in tempered tuning.

The resulting chord Scriabin considers to be a consonance. In fact it is an extension of the customary concept of a consonant chord, i.e., a chord that does not require a resolution. [61]

Our customary triad is only one example of this chord, an example in which some sounds are omitted:

–D–F-sharp–A– (major)

(C E B-flat omitted)

–A–C–E– (minor)

(B-flat D F-sharp omitted)

–F-sharp–A–C (diminished)

(B-flat D E omitted)

–D–F-sharp–B-flat (augmented)

(C E A omitted)

Even without context and development, we find a strange "mystical" atmosphere in this single harmony, something that reminds one of a deep-sounding enormous bell; if played in a high pitch, it shines, radiates, irritates, elevates, agitates. This harmony includes a much greater variety than the customary triad, which is only one case of this harmony. It must be remarked that this variety is not in the least exhausted in *Prometheus*. Here Scriabin uses almost entirely this one harmonic principle, which causes a strange impression. To the listener who becomes absorbed in the world of these harmonies and feels its "consonance," the whole texture of *Prometheus* becomes highly transparent. *Prometheus* is infinitely simple and entirely "consonant" so that *not a single dissonance* is to be found in it. This may also be explained by the fact that due to a great number of sounds in this harmony the composer can avoid all the changing and passing tones that are not included within it. All melodic voices are built up on the sounds of the accompanying harmony, all counterpoints are subordinated under the same principle. [62]

It is this fact only—in spite of the complete "consonance" and the utter transparency of the opus—that makes it possible to have five or six different themes played at the same time and to unite the thematic origin of the figures. In the whole of world literature *Prometheus* is the most complicated polyphonic work and, at the same time, in its texture, the most transparent.

It is not uninteresting to trace the evolution of Scriabin's harmonies from his earliest works on.

Already in his Waltz op. 1 (published by Jürgenson) we find the harmony:

A-flat F-flat (A-flat) C G-flat (G-flat C F-flat A-flat)

Without difficulty we recognize traits of the future ecstatic Scriabin. Only two tones that occur in the harmony of *Prometheus* are absent: B-flat and E-flat.

After a rather long time, at the time of the Second Symphony and of the Third Sonata, these harmonies reappeared, but not yet complete, i.e., not yet in the form of the so-called ninth chord with an augmented (or diminished) fifth. In this form Scriabin's chord enters the whole tone scale, even if its organic origin is still far away from this scale.

By the time Scriabin wrote his Third Symphony, this chord had become "dominant" in his work. It was the time when he wrote forty small works during one summer, including the *Tragic Poem* and the *Satanic Poem*, the *Poems op. 32*, the Fourth Sonata. Here for the first time the harmony of *Prometheus* appears in full, e.g., in Preludes op. 37 no. 2 (the sixth measure):

G-sharp F-sharp A-sharp B-sharp E-sharp C-double-sharp

E. Nolde

This complete form is not to be found very often during this period. It appears more frequently during the last phase (*Poem of Ecstasy*, Fifth Sonata).

In the *Poem of Ecstasy* the synthetic harmony appears at the culmination point (on page 41 of the score):

E-flat A F-sharp C G B-flat (E-flat G)

In the Fifth Sonata, which is harmonically closer to *Prometheus* than to the *Poem of Ecstasy*, we find it in the second theme; in *Fragilité* and other small works of the period it appears very frequently. But its consistent and complete execution we observe only in *Prometheus.*

With it begins the *Prometheus* poem of the creative spirit which, having already become free, freely creates the world. This is a kind of symphonic synopsis of the mystery, in which the participants are forced to experience the whole evolution of the creative spirit, [63] where they are separated into receiving, passive, and creative human interpreters. This separation is also to be seen in *Prometheus*: it has the customary form of the symphony, performed by orchestra and chorus.

In a blue-lilac twilight the mystical harmony sounds; when it flutters the main theme (1) is sounded by the French horns:

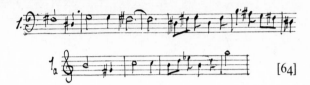

[64]

During these sounds the grandiose idea of the original chaos arises, in which for the first time the will of the creative spirit becomes audible (theme 2):

I have written out below the main themes of *Prometheus* from which the composer creates the texture of his orchestral music.

[65]

137

[66]

Study for *Composition No. 4* by Kandinsky [In color in the original edition]

His "single" harmony has the capacity to include the most diverse nuances, beginning with a mystical horror and ending with a radiant ecstasy and a caressing eroticism. I insist that *never before* in music has such a horror been heard as occurs in the tragic passages of *Prometheus*, that in no work has such an enthusiastic exaltation been heard as occurs shortly before the end; [67] an exaltation beside which the conclusion of the *Ecstasy* pales. According to the composer's idea the whole hall is filled with blinding rays at the same time that all the forces of the orchestra and the chorus are mobilized and the main theme is played by the trumpets against the background of broad orchestral and organ harmonies.

After this grand exaltation there is a sudden silence; the lights go off. In a lilac twilight the sounds of an ecstatic, intoxicating dance are heard. There are light effects, magic plays of sound, dashing "light-bearing" passages of the piano against the background of hissing cymbals. Another exaltation, and again the orchestra is caught up in a sea of sounds that merge in the final chord. This chord is the only "triad" the composer uses in his entire work. [68]

# Free Music
*by N. Kulbin*

## Theses of Free Music

The music of nature is free in its choice of notes—light, thunder, the whistling of wind, the rippling of water, the singing of birds. The nightingale sings not only the notes of contemporary music, but the notes of all the music it likes.

Free music follows the same laws of nature as do the music and the whole art of nature.

Like the nightingale, the artist of free music is not restricted by tones and half tones. He also uses quarter tones and eighth tones and music with a free choice of tones.

This disturbs neither the simplicity nor the search for a basic character, nor does it lead to a photographic reproduction of life, but it facilitates stylization.

In the beginning quarter tones are introduced. (In antiquity, when man was still rich in original instincts, they were used as the "enharmonic scale." They still exist in the old Hindu music.) [69]

M. Pechstein

## The Advantage of Free Music

New enjoyment of unusual tone combinations.
New harmony with new chords.
New dissonances with new resolutions.
New melodies.
The choice of possible chords and melodies is very much enlarged.

The power of musical poetry is magnified. This is most important because music is mainly poetry. Free music has many more possibilities for affecting the listener and exciting his soul.

Delicate combinations and changing tones strongly affect man's soul. [70]

The representational capacity of music is enhanced. The voice of a loved person can be rendered; the singing of the nightingale, the rustling of leaves, the delicate and stormy noise of the wind and the sea can be imitated. The movement of man's soul can be represented more completely.

142

Le Fauconnier: *Abundance*

W. Morgner

Study and use of colored music are facilitated.

A simple, effective method of training and developing hearing is maintained. Such exercises are necessary for the student.

A series of still unknown phenomena is revealed: *the close connection of tones and the processes of close connection.*

These connections of adjunct tones of a scale, of quarter tones or even lesser intervals, may still be called close dissonances, but they possess special characteristics that customary dissonances do not.

These close connections of tones evoke unusual sensations in man.

The vibration of closely connected tones is extremely exciting.

In such processes the irregular beat and the interference of tones (which is similar to that of light) are of great significance.

The vibration of close connections, their unfolding, their manifold play, make the representation of light, colors, and everything living much more effective than customary music does. It is also easier to achieve a lyrical atmosphere.

These close connections also create musical paintings, which consist of special planes of color that merge to form progressing harmonies, similar to contemporary painting.

## The Music of Free Tones

Progress in music is possible when the artist is not bound to notes, when he can use any interval, e.g., a third tone or a thirteenth tone. [71]

This music provides full freedom to inspiration, and it has the previously mentioned advantages of natural music: it can represent subjective experiences and, at the same time, the poetry of moods and passions as well as the illusions of nature.

## Practical Performance of Free Music

The listeners:

Those who think that quarter tones are hard to distinguish are wrong. We know from experience that all listeners can easily differentiate between quarter tones.

Eighth tones are not differentiated by all listeners. Their effect is stronger because these half-realized, incomprehensible sensations strongly affect man's soul.

O. Müller

The performance:

The performance of free music is very simple. As works with quarter tones can be performed, the improvisation of free tones by singing, by playing on the double bass, the cello, and some wind instruments can also be accomplished without any change and without a different tuning.

The harp can be tuned to quarter and other tones. The "chromatic" harp is best for this purpose.

The finger boards of the guitar, the zither, the balalaika, etc. must be changed.

The piano can also be tuned differently, but the number of octaves will be reduced and the keyboard will lose its significance. To avoid this, two rows of strings and two keyboards could be arranged.

Other instruments are also easy to use and adapt.

The simplest way to study the characteristics of free music is to use glass bowls or glasses and to fill them with different amounts of water.

It is also easy to build xylophones at home.

The writing of free music:

The staff remains almost unchanged. At the beginning it will only be necessary to add symbols for the quarter tones.

The improvisation of free tones may for the time being be taken down on Gramophone records.

They may also be depicted in the form of a drawing with rising and falling lines. [73]

# On the Question of Form
*by Kandinsky*

At a certain time what is inevitable ripens, i.e., the creative *spirit* (which could be called the abstract spirit) makes contact with the soul, later with other souls, and awakens a yearning, an inner urge.

When the conditions necessary for the maturation of a certain form are met, the yearning, the inner urge, the force is strengthened so that it can create a new value in the human spirit that consciously or unconsciously begins to live in man.

Consciously or unconsciously man tries, from this moment on, to find a material form for the spiritual form, for the new value that lives within him.

This is the search by the spiritual value for materialization. Matter is a kind of larder from which the spirit chooses what is *necessary* for itself, much as a cook would.

This is the positive, the creative. This is goodness. *The white, fertilizing ray.*

This white ray leads to evolution, to elevation. Behind matter, within matter, the creative spirit is hidden.

The veiling of the spirit in matter is often so thick that, generally, only a few people can see through it to the spirit. There are many people who cannot even recognize the spirit in a spiritual form. Today many do not see the spirit in religion, in art. There are whole epochs that deny the spirit, because the eyes of man cannot see the spirit at

147

Votive Painting

those times. So it was during the nineteenth century and so it is for the most part today.

Men are blinded.

A black hand covers their eyes. The black hand belongs to the hater. The hater tries with every available means to slow down the evolution, the elevation.

This is negative, this is destructive. This is evil. *The black, fatal hand.* [74]

Evolution, movement forward and upward, is possible only when the way is clear, free of obstacles. This is the *outer condition.*

The force that propels the human spirit on the clear way forward and upward is the abstract spirit. It must be audible and it must be heard. The call must be possible. *This is the inner condition.*

To destroy these two conditions is the method the black hand uses against evolution.

Its tools are: fear of the clear way, of freedom (Philistinism), and deafness to the spirit (dull materialism).

That is why men regard each new value as inimical. They try to fight against it with scorn and defamation. The bearer of values is regarded as ridiculous and dishonest. They laugh at and scorn the new value.

That is the horror of life.

The joy of life is the incessant, constant victory of the new value.

This victory proceeds slowly. Gradually the new value conquers man. And when many men no longer question this value, indispensable and necessary today, then it will form a wall erected against tomorrow.

The metamorphosis of the new value (the fruit of freedom) into a fossilized form (a wall against freedom) is the work of the black hand.

All evolution, which is internal development and external culture, is thus a shifting of obstacles.

Obstacles destroy freedom. By this destruction they prevent us from hearing the new revelation of the spirit.

Obstacles are constantly created from new values that have pushed aside old obstacles.

We see that basically the most important thing is not the new value, but rather the spirit that has revealed itself in the new value. And the freedom necessary for the revelations.

We see that the absolute cannot be sought in the form (materialism).

Form is always temporal, i.e., relative, for it is nothing more than the means necessary today through which the present revelation makes itself heard.

Sound, therefore, is the soul of form, which only comes alive through sound and which works from the inside out.

*Form is the outer expression of the inner content.* [75]

We should never make a god out of form. We should struggle for form only as long as it serves as a means of expression for the inner sound. Therefore we should not look for salvation in *one* form only.

149

"Sitting"

This statement must be understood correctly. For each artist (i.e., creative artist, not interpreter) his means of expression (= form) is the best, since it embodies best what he is compelled to reveal. From this principle people often draw the wrong conclusion: that this means of expression is or should be the best for other artists as well. [76]

Since form is only an expression of content, and content is different with different artists, it is clear that there may be *many different forms at the same time* that are *equally good*.

*Necessity creates form.* Fish that live at great depths have no eyes. The elephant has a trunk. The chameleon changes its color, etc., etc.

Form reflects the spirit of the individual artist. Form bears the stamp of the *personality*.

The personality cannot, of course, be regarded as something outside of time and space. It depends to a certain extent on time (epoch) and space (people).

As each individual artist has to express himself, so each people has to express itself, including the people to which the artist belongs. This connection is reflected in the form and is described as the *national element*.

And, finally, each period has its special task as well, the revelation possible through it. The reflection of this time element is recognized in the work as *style*.

All these three elements leave their stamp on a work and are unavoidable. It is not only superfluous to worry about their existence, but also harmful. Forcing them would be pretending, betraying the period.

On the other hand, it is obvious that it would be superfluous and harmful to emphasize only one of these three elements. As many today emphasize the national element, and still others emphasize style, people recently have paid homage to the cult of the personality (the individual element).

As we said at the beginning, the abstract spirit first overtakes a single human spirit; later it controls an increasing number of people. At that moment some artists succumb to the spirit of a time, which forces them to adopt some forms that are related to each other and therefore share an external similarity.

This moment is called a *movement*.

The movement is completely justified and essential for a group of artists (as is the individual form for a single artist). [77]

151

Henri Rousseau

As one cannot find salvation in the form of an individual artist, so one cannot find it in the form of the group. For each group its form is the best because it best embodies what the group is compelled to reveal. From this we should not conclude that this form is or should be the best for all. Here, too, there should be complete freedom. Each form that is the external expression of the internal content should be considered valid, and each form should be considered genuine (= artistic). If one acts differently, one no longer serves the free spirit (white ray) but the fossilized obstacle (black hand).

Thus we come to the same conclusion that was stated above: it is not form (matter) that is generally most important, but content (spirit).

Thus form may appear as pleasant, unpleasant, attractive, unattractive, harmonious, disharmonious, skillful, unskillful, delicate, coarse, and so on. But it must not be accepted or rejected for qualities considered to be either positive or negative. All these concepts are completely relative, as can be seen in a quick glance at the infinitely changing series of forms that have already existed.

Form itself is just as relative and should be valued and considered as such. We should approach a work so that its form affects the soul and through the form its content (spirit, interior sound). Otherwise we elevate the relative to the absolute.

In daily life we would rarely find a man who will get off the train at Regensburg when he wants to go to Berlin. In spiritual life, getting off at Regensburg is a rather common occurrence. Sometimes even the engineer does not want to go on, and all the passengers get off at Regensburg. How many who sought God stopped at a carved figure! How many who searched for art were arrested at a form that an artist had used for his own purposes, be it Giotto, Raphael, Dürer, or Van Gogh!

It must be stated as a final conclusion: the most important thing is not whether the form is personal, national, or has style, whether or not it corresponds to a major contemporary movement, whether or not it is related to many or few other forms, whether it is unique, etc., etc. *The most important thing in the question of form is whether or not the form has grown out of inner necessity.*[1] [78] The existence of the forms in time and space can be explained as arising out of the inner necessity of time and space.

Therefore, it will finally be possible to penetrate to the characteristics of a period and a people and to represent them systematically.

The greater the epoch, the greater (quantitatively and qualitatively) the striving toward the spiritual, the more various the forms will be,

---

[1]This means that one should not make a uniform out of the form. Works of art are not soldiers. One and the same form can therefore, even with the same artist, be at one time the best, at another the worst. In the first case it grew in the soil of inner necessity, in the second in the soil of outer necessity: out of ambition and greed.

Drawings by Children

and the more numerous the collective currents (group movements) to be observed. This is obvious.

These characteristics of a great spiritual epoch (which was prophesied and is today in its initial stage) can be seen in contemporary art. They are:

1. a great *freedom*, which appears to some to be limitless and which
2. makes the *spirit* audible, which
3. we see revealing itself with an overwhelming *force* in things that
4. will increasingly and already do make use of all *spiritual fields*, so that
5. in each spiritual field, including the plastic arts (especially painting), it will create  many *means of expression* (forms) encompassing both individuals and groups, for
6. today it has the whole larder at its disposal, i.e., *every* material, from the most "solid" to that which exists only in two dimensions (abstract), will be used as an element of form.

Elaboration to 1: As far as freedom is concerned, it expresses itself in the effort toward liberation from forms that have already reached their fulfillment, i.e., liberation from old forms in the effort to create new and infinitely varied forms.

Elaboration to 2: The automatic search for the outermost limits of our present means of expression (of the personality, the people, the time) is, on the other hand, a limitation of apparently unrestrained freedom, which is determined by the spirit of the time, and a clarification of the direction that the search must take. A little insect under glass, running in all directions, thinks it has unrestrained freedom. At a certain point it hits the glass: it can see farther but it cannot go any farther. When you move the glass forward, you give it the potential of covering more space. But its main movement is determined by the guiding hand. Similarly, our time believes itself completely free, but we will encounter certain limits, which will, however, be rearranged "tomorrow." [79]

154

Bavarian Glass Painting

Arnold Schönberg

Elaboration to 3: This apparently unbridled freedom and the intervention of the spirit arise from the fact that we begin to feel the spirit, the *inner sound*, in all things. And at the same time this emerging ability is the ripening fruit of the apparently unbridled freedom and the active spirit.

Elaboration to 4: We cannot specify here the above-mentioned effects on all other spiritual disciplines. But it should be clear to everyone that the cooperation of freedom and spirit will sooner or later be reflected everywhere.[2]

Elaboration to 5: In the plastic arts (more especially in painting) we encounter today a striking wealth of forms, which seem to be partly the forms of great individual personalities, partly of whole groups of artists, and which are swept along in a great, well-defined flowing current.

Yet in the great variety of forms the common effort is easily recognized. And in just this same mass movement the all-embracing spirit of form is recognized today. It is enough to say: *everything is permitted.* What is permitted today cannot be transgressed. What is forbidden today remains unshaken. [80]

One should not set up limits because they exist anyway. This is true not only for the sender (artist) but also for the receiver (viewer). The viewer can and must follow the artist, and he should not be afraid of being misguided. Man cannot move in a straight line physically (look at the paths in fields and meadows!), much less spiritually. And on spiritual paths especially, the straight line is often the longest because it is false, and the apparently false path is often the right one.

The "feeling" that speaks aloud will sooner or later correctly guide the artist as well as the viewer. The fearful clinging to a *single* form will inevitably lead ultimately to a dead end. Open feeling—to freedom. The first follows matter. The second—the spirit: the spirit creates a form and goes on to other forms.

Elaboration to 6: The eye directed toward one point (either form or content) cannot possibly view a wide plane. The unobservant eye glancing over the whole surface sees the wide plane, or a part of it, but becomes caught up in external differences and gets lost in contradictions. These contradictions arise because of the variety of means that the contemporary spirit draws out of its larder of matter, apparently without plan. Many call the present state of painting "anarchy." The same word is also used occasionally to describe the present state of music. It is thought, incorrectly, to mean unplanned upheaval and disorder. But anarchy is regularity and order created not by an external and ultimately powerless force, but by *the feeling for the good.* Limits are set up here, too, but they must be *internal* limits and must replace external ones. These limits are also constantly extended, giving rise

[2]I have elaborated on this point in my essay *Concerning the Spiritual in Art.*

157

Henri Rousseau

to an ever-increasing freedom that, in turn, opens the way for subsequent revelations. [81]

Contemporary art in this sense is truly anarchistic: it not only reflects the spiritual standpoint already conquered but also embodies the spirit as a materializing force, ripe for revelation.

These two elements have always existed in art. They have been matter by the spirit can easily be classified between two poles.

These two poles are:

1. Total abstraction
2. Total realism

These two poles open *two ways* that lead ultimately *to one goal.*

Between these poles lie many combinations of the different harmonies of abstraction and realism.

These two elements nave always existed in art. They have been classified as the "purely artistic" and the "objective." The first was expressed in the second while the second was serving the first. It was a fluid balance, which seemed to search for its ideal fruition in an absolute equilibrium.

It seems that this ideal is no longer a goal today. The horizontal bar that held the two pans of the scale in balance seems to have vanished

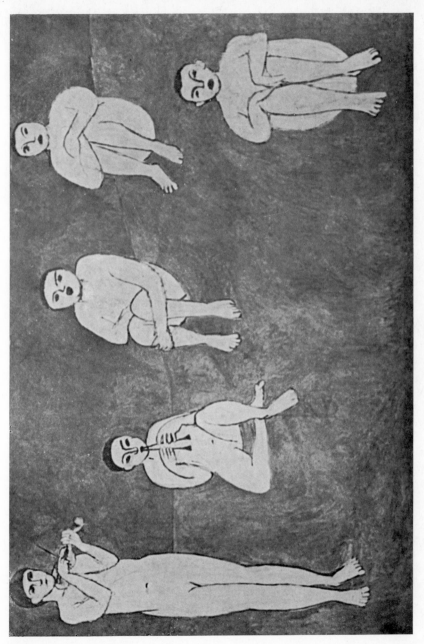

Henri Matisse: *La Musique*

Votive Painting

today; each pan intends to exist individually and independently. This breaking of the ideal scale also seems to people to be "anarchistic." Art has apparently brought to an end the pleasant supplementing of the objective by the abstract, and vice versa.

On the one hand, the diverting support of reality has been removed from the abstract, and viewers think they are floating. They say that art has lost its footing. On the other hand, the diverting idealization of the abstract (the "artistic" element) has been removed from the objective; and viewers feel nailed to the floor. They say that art has lost the ideal.

These reproaches originate in incompletely developed feeling. The

160

Votive Painting

custom of concentrating on form and the resultant behavior of the viewer who clings to the accustomed form of balance are the blinding forces that block free access to free feeling. [82]

The emergent great realism is an effort to banish external artistic elements from painting and to embody the content of the work in a simple ("inartistic") representation of the simple solid object. Thus interpreted and fixed in the painting, the outer shell of the object and the simultaneous canceling of conventional and obtrusive beauty best reveal the inner sound of the thing. With the "artistic" reduced to a minimum, the soul of the object can be heard at its strongest through

161

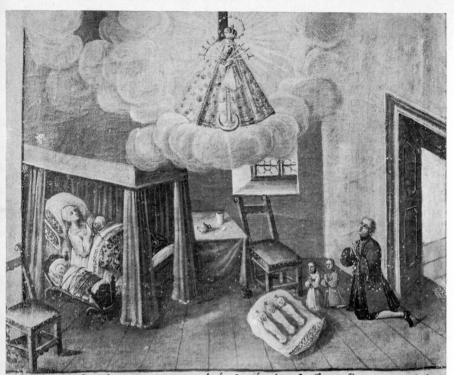

Fraü Maria Schmidin von Isfingen wurde, da sie in den 6 Wochen Tag, gechling mit einem solchen schwermuett vnd Gröchen iber fallen, daß sie gantz von verstandt komen: auch nicht fehig warr daß H: Sacramenta zü Empfangen. nach dem Ihr Eheman sie selbe mit einem Lobambt Votivtafel vnd Opfer in soch verlobt. so hat obermannte alsogleich ihren verstand in kürtzer Zeit, darauf aber die völlige gesundheit erhalten
1763

Votive Painting

its shell because tasteful outer beauty can no longer be a distraction.[3] [83]

This is possible only because we can increasingly hear the whole world, not in a beautiful interpretation, but as it is.

*The "artistic" reduced to a minimum must be considered as the most intensely effective abstraction.*[4]

[3]The spirit has already absorbed the content of conventional beauty and no longer finds new nourishment in it. The form of this conventional beauty gives conventional enjoyment to the lazy physical eye. The effect of the work is mired in the field of the physical. Spiritual experience becomes impossible. Often this beauty creates a force that leads not to the spirit but away from it.

[4] *The quantitative reduction of the abstract therefore equals the qualitative intensifica-*

Franc: Xaveri Xottmiller Müller zu Mühlhagen: Lage gefärlich Xrank an der roffen ruße, nicht minder
stünde er in gefahr eines seiner besten Pferdm zu verliehren, seine grofan also anbelangenheid verlobt er
sich anhero mit einer H: Mef auch Opfer in Stock: von dem Pferdt aber verspricht er den halben Wert
in dem Stock zu legen. Wo er auch in beden anligenheiten recht augenscheinliche hilf erlanget.
den. zw. Aug. 1766

Votive Painting

*tion of the abstract.* Here we touch one of the most essential rules: the *external*
enlargement of a means of expression leads under certain circumstances to a
reduction of its *internal* power. Here 2 + 1 is less than 2 − 1. This rule is naturally
also revealed in the smallest forms of expression: a spot of color often loses intensity and must lose effect by being enlarged and made more powerful. An especially
active movement of color is often produced through restraint; a mournful sound
can be achieved by direct sweetness of color, and so on. All these are expressions of
the law of antithesis and its consequences. In short: *true form is produced from the
combination of feeling and science.* Again I must remind you of the cook! A good
substantial meal is produced from a combination of a good recipe (where everything
is specified in pounds and ounces) and guiding feeling. An important characteristic
of our time is the increase of knowledge: aesthetics gradually assumes its proper
place. This is the future "thorough bass," ahead of which naturally lies infinite
change and development!

163

Henri Rousseau

The great antithesis to this realism is the great abstraction, which apparently intends to annihilate the objective (reality) and to embody the content of the work in "incorporeal" forms. Thus interpreted and fixed in the painting, the asbtract life of representational forms is reduced to a minimum and best reveals the inner sound of the painting. Likewise, as in realism the inner sound is intensified by blotting out the abstract, so in abstraction this sound is intensified by blotting out reality. In realism conventional outer, tasteful beauty is a limitation; in abstraction the conventional, outer, supporting object is a limitation.

In order to "understand" this kind of painting, the same kind of liberation is necessary as in realism. That is, here also it must become possible to hear the whole world as it is without representational interpretation. Here these abstracting or abstract forms (lines, planes, dots, etc.) are not important in themselves, but only their inner sound, their life. As in realism, the object itself or its outer shell is not important, only its inner sound, its life.

*The "representational" reduced to a minimum must in abstraction be regarded as the most intensely effective reality.*[5]

In conclusion: in total realism the real appears strikingly great and the abstract strikingly small; in total abstraction this relation [84] seems to be reversed, so that in the end these (= aim) two poles are equalized. Between these two antipodes we can put the equals sign:

Realism = Abstraction

Abstraction = Realism

*The greatest external difference becomes the greatest internal equality.*

A few examples will lead us from the theoretical sphere to the practical. When a reader looks at some letter in these lines with unskilled eyes, he will see it not as a familiar symbol for a part of a word but first as a *thing*. Besides the practical man-made abstract form, which is a fixed symbol for a specific sound, [85] he will also see a physical form that quite autonomously causes a certain outer and inner impression; it is independent of the above-mentioned abstract form. In this sense the letter consists of:

Element 1. The main form = the overall appearance, which, roughly speaking, denotes "happy," "sad," "striving," "sinking," "defiant," "ostentatious," etc., etc.;

Element 2. Various specifically curved lines, which always make a certain internal impression; they too can be "happy," "sad," etc.

When the reader has felt these two elements of the letter, a feeling is immediately aroused in him caused by this letter as a *being* with *inner life*.

One should not reply that this letter affects one person in one way, and another in another way. This is irrelevant and understandable. Generally speaking, every being affects different persons differently. We see only that the letter consists of two elements that in the end express *one* sound. The various lines of the second element may be "happy," but the overall impression (Element 1) may still be "sad." The various movements of the second element are organic parts of the first. The same construction and the same subordination of the various elements under *one* sound may be observed in any lied, any piano piece, any symphony. And exactly the same procedure forms a

[5]At the other pole we meet the same previously mentioned law whereby *quantitative reduction equals qualitative intensification.*

Arnold Schönberg

drawing, a sketch, a painting. Here the rules of construction are revealed. At the moment only one thing is important to us: the letter produces an effect. As mentioned before, this effect is twofold:

1. The letter acts as a purposeful symbol;
2. It first acts as form and later as the inner sound of this form, self-supporting and entirely independent.

It is important to us that these two effects are not connected with each other; while the first effect is external, the second has an inner meaning.

Unknown Master

From this we conclude that the *outer effect can be different from the inner*, which is produced by the *inner sound*. This is *one of the most powerful* and most profound *means of expression* in any composition.[6] [86]

Let us take another example. Suppose we see a dash in the same book. If this dash is in its proper place—as I use it here—it is a line with a practical purpose. Let us lengthen this little line and still leave it in its proper place: the meaning of the line remains the same, but by its unusual length it is given an undefinable shading. The reader asks himself why this line is so long and whether this length has a practical purpose. Let us put the same line in a wrong place (as—I do here). The practical function is lost and cannot be found; the shade of the

[6]Here I can only touch in passing on such great problems. If the reader wishes to become absorbed in these questions, he will realize the power, the irresistibly alluring mystery, for example, of this last chain of reasoning.

question has increased. It might be a misprint, i.e., a distorted practical purpose. This sounds negative. Let us draw the same line on an empty page, long and curved. As with the last example, one expects to learn its real practical function (as long as there is hope for an explanation). Afterward (when there is no explanation) one thinks of its negativeness.

As long as this or that line remains in a book, its practical purpose cannot definitely be eliminated. [87]

Let us draw a similar line in a context where practical purpose can be completely eliminated, e.g., on a canvas. As long as the viewer (he is no longer a reader) looks at this line on the canvas as a means of delineating an object, he is still subject to the impression of the practical purpose. But at the moment he sees that the practical object in the painting is mostly accidental and does not play a purely pictorial role, and that the line can sometimes have a purely pictorial significance,[7] at that moment the viewer's soul is mature enough to perceive the *pure inner sound* of this line.

Is the object, the thing, thus eliminated from the painting?

No. As we saw above, the line is a thing, which has the same practical purpose as a chair, a fountain, a knife, a book, and so on. And this thing is used in the last example as a purely pictorial means, without the other aspects that it otherwise may have—that is, in its pure inner sound. In a painting, when a line is freed from delineating a thing and functions as a thing in itself, its inner sound is not weakened by minor functions, and it receives its full inner power.

We may conclude that pure abstraction makes use of things that lead a material existence just as pure realism does. Again the complete negation of objects is equal to their complete affirmation. In both cases the equals sign is again justified because it leads to the same end: embodiment of the selfsame inner sound.

Here we see that in principle it makes *no difference whether the artist uses real or abstract forms.*

*Both forms are basically internally equal.* The choice must be left to the artist, who must know best by which means he most clearly can give material expression to the content of his art.

To express it abstractly: *in principle there is no question of form.* [88]

If there were in fact a question of form in principle, an answer would

[7]Van Gogh utilized the line as such with special power, without thereby intending to denote the corporeal.

Henri Rousseau

be possible. Everyone who had the answer would be able to create works of art, i.e., art would at that very moment cease to exist. Practically speaking, the question of form becomes the question: which form should I use in this case in order to achieve the necessary expression of my inner experience? The answer is scientifically precise and absolute in this case and relative in other cases, i.e., a form that is the best in one case may be the worst in another. Everything depends on the inner necessity, which alone can determine the appropriate form. And one form may be appropriate for several artists only when, under the pressure of time and space, the inner necessity chooses several related forms. However, this does not change the relative meaning of form at all, because an appropriate form in one case may still be inappropriate in other cases. [89]

All the rules discovered in earlier art and those to be discovered later —which art historians value too highly—are not general rules: they

Egyptian

do not lead to art. If I know the craft of carpentry, I will always be able to make a table. But one who knows the supposed rules of painting will never be sure of creating a work of art.

These supposed rules, which will soon lead to a "thorough bass" in painting, are merely the recognition of the inner effect of various methods and their combination. But there is no rule by which one can arrive at the application of effective form and the combination of particular methods precisely necessary in a specific case.

The practical result: *one should never trust a theoretician (art historian, critic, etc.) who asserts that he has discovered some objective mistake in a work.*

*The only thing* a theoretician is justified in asserting is that he does not yet know this or that method. If in praising or condemning a work theoreticians start from an analysis of already existing forms, they are most dangerously misleading and create a wall between the work and the naive viewer.

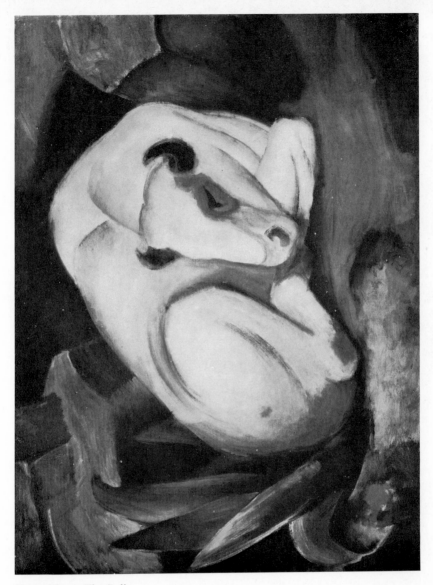

Franz Marc: *The Bull*

Japanese

From this point of view (which unfortunately is usually the only possible one) *art criticism is the worst enemy of art.*

*The ideal art critic* would not want to discover "mistakes,"[8] "errors," "ignorance," "plagiarism," etc., etc. Rather, he would try *to feel* how this or that form works internally, and then he would convey his total experience vividly to the public.

Such a critic would need the soul of a poet, because the poet has to feel objectively in order to express his feeling subjectively. This means that the critic would need creative ability. In reality, critics are very often unsuccessful artists, who are frustrated by the lack of creative ability of their own, and therefore feel called upon to guide the creative ability of others.

Another reason why the question of form is often harmful to art is that untalented people (i.e., people without an *inner* urge for art) use forms not their own, thus faking works and causing confusion. [90]

[8]E.g., "anatomical mistakes," "distortions," etc., or offenses against the future "thorough bass."

I must be precise here. It is considered a crime and fraud by critics, by the public, and often by the artist, to use the forms of others. This, in fact, is only true when the "artist" uses these forms without an inner necessity and creates a lifeless, dead, spurious work. If in order to express his inner impulses and experiences, an artist uses one or another "borrowed" form in accord with inner truth, he then exercises his right to use every form *internally necessary* to him—a utilitarian object, a heavenly body, or an artistic form created by another artist.

The whole question of "imitation"[9] is also far from having the importance attached to it by the critics.[10] The living remain. The dead vanish. [91]

Really, the further we look back into the past, the fewer faked and spurious works we find. They have vanished mysteriously. Only the genuine creation remains; i.e., only that which has a soul (content) in its body (form).

If the reader looks at any object on his table (even if it is only a cigarette stub) he will immediately observe the same two effects. Whenever and wherever it is (in the street, in church, in the sky, in the water, in a stable, or in the woods), the same two effects will always emerge, and always the inner sound will be independent of the outer meaning.

*The world sounds. It is a cosmos of spiritually effective beings. Even dead matter is living spirit.*

---

[9]Every artist knows how fanciful the critics are in this field. They know that the most extravagant statements may be used here without consequences. For example, recently the *Negress* by Eugen Kahler, a good naturalistic studio painting, was compared with . . . Gauguin! The only reason for this comparison could be the brown skin of the model (see *Münchner Neueste Nachrichten*, October 12, 1911), etc., etc.

[10]Thanks to the prevailing exaggeration of this question the artist is discredited with impunity.

Henri
Rousseau

If we draw the necessary conclusions from the independent effect of the inner sound, we see that this inner sound increases in intensity if we remove its stifling, external, practical meaning. This explains the pronounced effect of a child's drawing on the impartial, untraditional viewer. The child is indifferent to practical meanings since he looks at everything with fresh eyes, and he still has the natural ability to absorb the thing as such. Only later does the child by many, often sad, experiences slowly learn about the practical meanings. Without exception, in each child's drawing the inner sound of the subject is revealed automatically. Adults, especially teachers, try to force the practical meaning upon the child. They criticize the child's drawing

Henri Rousseau

from this superficial point of view: "Your man cannot walk because he has only one leg," "Nobody can sit on your chair because it is lopsided," and so on.[11] The child laughs at himself. But he should cry.

Besides being able to ignore the external, the gifted child also has the ability to express the abiding internal in such a form that it emerges and affects very forcefully (or as people say, "It speaks"!). [92]

Every form has many aspects, and one may discover new and different virtues in it over and over again. But here I want to emphasize one aspect of good children's drawing important to us at the moment: composition. Here we see the evidently unconscious application—

[11]As is so often the case: those who should teach are taught. And later people wonder that these talented children do not develop.

virtually self-generated—of both previously discussed aspects of the letter: (1) the *total appearance*, which very often is precise and sometimes schematic, and (2) the *single forms*, which together create the great form, each leading a life of its own (e.g., *Arabs* by Lydia Wieber). There is an enormous unconscious power in the child that expresses itself here and that raises his work to the level of adults' work (sometimes even higher!). [12]

For every fire there is a cooling off. For every early bud—the threatening frost. For every young talent—an academy. These are not tragic words but a melancholy fact. The academy is the surest way of destroying the power of the child. Even the greatest, strongest talent is more or less retarded in this respect by the academy. Lesser talents perish by the hundreds. An academically trained person of average talent excels in learning practical meanings and losing the ability to hear his inner sound. He produces a "correct" drawing that is dead.

If a man without academic training, free of objective artistic knowledge, paints something, he never produces an empty sham. This fact demonstrates the effect of the inner power, which is influenced only by the *general* knowledge of the practical meaning.

However, only a limited application of this general knowledge is possible in these works. Here the external is removed (less so than with a child, but still considerably), and the inner sound is intensified: the works produced are not dead, but living (see, for example, the four heads reproduced here). [93]

[12] "Folk art" also possesses this startling quality of compositional form (see the votive painting of the Plague in the church in Murnau).

176

Alfred Kubin: Pen Drawing

Christ said: Suffer the little children to come unto me, and forbid them not; for of such is the kingdom of God.

The artist, whose whole life is similar in many ways to that of a child, can often realize the inner sound of things more easily than anyone else. In this respect it is extremely interesting to see how the composer Arnold Schönberg simply and confidently applies pictorial methods. As a rule he is interested only in the inner sound. He disregards decorations and delicacies entirely, and in his hand the "poorest" form becomes the richest (see his *Self-Portrait*).

Here lies the root of the new total realism. In rendering the shell of an object simply and completely, one has already separated the object from its practical meaning and peeled forth its interior sounds. Henri Rousseau, who may be considered the father of this realism, has pointed the way simply and convincingly (see the portrait and his other paintings).[13] [94]

Henri Rousseau has revealed the new possibilities of simplicity. At present this aspect of his complex talent is most valuable to us.

Objects or the object (i.e., the object and the parts forming it) must be in a certain relationship. This relationship may be strikingly harmonious or strikingly disharmonious. A schematic or a subtle rhythm may be applied.

The irresistible effort of artists today to expose the purely compositional and to reveal the future rules of our great epoch is what forces them to strive along different paths to the same goal. [95]

In such a case, the artist naturally turns to the most regular and at the same time to the most abstract. We see that in many artistic periods the triangle was used as a basis for construction. This triangle was often equilateral, and therefore the number became important; i.e., the completely abstract element of the triangle. In the present search for abstract relationships the number plays a particularly important role. Each numerical formula is cold, like an icy summit, and solid in its extreme regularity like a block of marble. It is cold and solid like all necessities. The search to express the compositional element in a formula is the origin of so-called cubism. This "mathematical" construction is a form that sometimes must lead to the complete destruction of the material relationship of the parts of the

[13]The majority of Rousseau's paintings reproduced here are taken from Uhde's warm, sympathetic book (*Henri Rousseau*, Paris, Eugène Figuière et Cie. Editeurs, 1911). I take this opportunity to thank Mr. Uhde cordially for his cooperation.

Egyptian Shadow Play Figure [In color in the original edition]

object, and when applied consistently, does so (see, for example, Picasso).

The final goal of this approach is to create a painting that becomes real, that comes alive, exists, through its own schematically constructed parts. If there is any objection to this approach, it is *only* that the number is too narrowly applied. Everything can be represented as a mathematical formula or simply as a number. But there are various numbers: 1 and 0.3333 . . . have equal rights and both are living beings with an inner sound. Why be satisfied with 1? Why exclude 0.3333 . . . ? This raises another question: Why restrict artistic expression to the *exclusive* use of the triangle and similar geometric forms and figures? It must be repeated that the compositional efforts of the "cubists" are directly connected to the necessity to create pure pictorial essence which, on the one hand, speaks in the object and through it and, on the other hand, in various combinations with the more or less resonant object, finally leads to pure abstraction. [96]

Between the purely abstract and the purely real composition lie possibilities of combining abstract and real elements in one picture. The reproductions in this book show how great and diverse the possibilities of combination are, how life pulsates strongly in all these examples, and how free one should be in considering the question of form.

The artist's inner intuition does and will always control the combination of the abstract and the objective, the choices among the infinite number of abstract forms or of objective matter, i.e., the choice of the particular means in both fields. Any form that today is in bad taste or despised, that seems to lie far off the beaten path, only awaits its master. This form is not dead, it is merely sunk in a kind of lethargy. When the content, the spirit, which can appear only through this seemingly dead form, becomes ripe, when its hour of materialization has arrived, the spirit enters this form and speaks through it.

The layman especially should not approach a work with the question: "What did the artist *not* do?" or else ask: "How does the artist dare to neglect *my* wishes?" Rather he should ask: "What did the artist do?" or "Which of *his* inner wishes did the artist express here?" I also believe that the time will yet come when the critics will no longer search for negative qualities, for mistakes, but will seek out and communicate positive appropriate qualities. One of the "important" problems for the contemporary critic of abstract art is how to differentiate between the false and the true, that is, for the most part,

G. Münter: *Still Life*

how to detect the negative qualities. The attitude toward a work of art should be different from the attitude toward a horse one wants to buy. With a horse one important negative quality outweighs all the positive ones and makes it worthless; with a work of art this relationship is reversed: one important positive quality outweighs all the negative ones and makes it valuable.

If this simple idea were taken into consideration, all questions of form as an absolute principle would vanish automatically. The question of form would retain its relative value, however, and the choice of form necessary for himself and for his work will finally be left to the artist. [97]

In closing these reflections on the question of form, which unfortunately are very short, I want to point out a few examples of construction reproduced in this book. I am forced to emphasize only one of the many vital aspects of the works and to omit all the other manifold qualities that characterize not just one specific work, but also the soul of the artist.

The two paintings by Henri Matisse show how the "rhythmic" composition (*La Danse*) has an internal life and consequently a sound that differs from the composition in which the elements of the painting seem to be combined unrhythmically (*La Musique*). This comparison is the best proof that harmony lies not only in a clear-cut scheme but also in a clear rhythmic pattern.

The strong abstract sound of physical forms does not necessarily demand the destruction of the objective. As we see in the picture by

*Foolish Virgin* (Thirteenth Century)

Oskar Kokoschka: *Portrait*

Henri Rousseau: *Portrait*

Marc (*The Bull*) there are also no general rules in this matter. The object can fully retain both inner and outer sound, and its individual parts may become independently sounding abstract forms, thereby causing a general abstract primary sound.

Münter's still life shows that a dissimilar and uneven transposition of objects in one and the same painting is not only harmless, but applied correctly, it produces a strong, complicated inner sound. The accord that seems to be disharmonious externally in this case creates the harmonious inner effect.

The two paintings by Le Fauconnier are powerful, didactic examples: similar "relieflike" forms cause two diametrically opposed inner effects by a distribution of "weights." *Abundance* resounds like an almost tragic overloading of weights. *Swamp* reminds one of pure lucid poetry. [98]

If the reader is able to rid himself of his own desires, his own ideas, his own feelings for a while and leafs through this book, going from a votive painting to Delaunay, from Cézanne to a work of Russian folk art, from a mask to Picasso, from a glass painting to Kubin, etc., etc., then his soul will experience many vibrations and he will enter the sphere of art. Here he will not find shocking defects and annoying faults, and instead of a minus he will attain a spiritual plus. And these vibrations and the plus arising from them will enrich his soul as no means other than art can do.

Later the reader can go on with the artist to objective reflections and scholarly analysis. Then he will find here that all the examples respond to the same inner call—composition; that they all restore the same inner basis—construction.

The inner content of a work belongs to one of two processes that include all the minor movements today (only today? or distinctly visible only today?).

These two processes are:

1. Disintegration of the soulless, materialistic life of the nineteenth century, i.e., the collapse of the material supports that were considered the only solid ones and the decay and dissolution of the various parts.

2. Construction of the spiritual and intellectual life of the twentieth century that we experience and that is already manifested and embodied in strong, expressive, and distinct forms.

These two processes are the two aspects of the "modern movement."

To define what has been achieved or what is the final goal of the movement would be presumptuous and would immediately be cruelly punished by a loss of freedom.

As we have often said before, we should strive not for restriction but for liberation. One should discard nothing without making an *extreme effort* to discover its living qualities. [99] It is better to think of death as life than to think of life as death—even just once. And only on a spot that has become free can something *grow*. The free man tries to enrich himself with everything and to let the life in every being affect him—even if it is only a burnt match.

*The future* can be received only through freedom.

One cannot stand apart like the barren tree under which Christ saw the sword ready and waiting. [100]

# From *Italian Impressions* by W. Rozanov[1]

In contrast to modern art the art of antiquity as a whole is not psychological. . . . But perhaps the art of antiquity was more metaphysical?

In all those marble works we find the dimensions again and again, the measurements of the human corpus, the unending search for (and perhaps discovery of?) the definitive truth of these dimensions and of their harmony. We could call this "tailor's calculations." Does this not seem to be very trifling, very poor? But what did Moses say when he came down from Mount Sinai, and what did he tell the Children of Israel about the building of the Temple (Tabernacle)? He only gave details about dimensions and colors, almost nothing but dimensions. When we read the account of the "Exodus of the Children of Israel" we can almost hear the tailor call out the figures for the garments ordered—length, width, girth, bending. The Tabernacle is God's

[1](St. Petersburg, 1909, pp. 81ff.)

187

Bavarian Mirror Painting

dress: this is their [101] unexpressed idea. The prophet Ezekiel says nothing about his impression of the envisioned Temple, nothing about where God is, nothing about the image of this Temple. But he tires and exhausts the reader's patience with pages of numbers and more numbers, with dimensions and more and more dimensions. And the sage Pythagoras considered the "number" to be the "essence of all things." "Each thing has its own number. He to whom the number of the thing has been revealed also knows the hidden essence of things." There is a specific mystery about numbers and measures; *God* is the measure of all things—*after* the Creation. *Before* the Creation, should he not be called the tailor of all things, "cutting" the world according to his heavenly meaning? [102]

St. Antonius de Padua.

# On Stage Composition
*by Kandinsky*

Each art has its own language, that is, its own methods.

Each art is something complete in itself. Each art leads a life of its own. It is an empire in itself. [103]

Therefore the methods of the various arts are completely different externally. Sound, color, word! . . .

*In their innermost core* these methods are wholly identical: their final goal obliterates external differences and reveals their inner identity.

The *final* goal (knowledge) is reached through delicate vibrations of the human soul. These delicate vibrations are ultimately identical, although their inner motions are different.

The undefinable and still distinct spiritual action (vibration) is the goal of the various methods of art.

A distinctive complex of vibrations is the goal of a work.

The refinement of the soul through the accumulation of distinctive complexes—this is the goal of art.

*Art is*, therefore, indispensable and *practical*.

When the artist finds the appropriate means, it is a material form of his soul's vibration, which he is forced to express. [104]

If the method is appropriate, it causes an almost identical vibration in the soul of the audience.

This is inevitable. But this secondary vibration is complicated. It may be strong or weak; this depends on the audience's level of development and on the influence of the times (the absorbed soul). Second, these vibrations in the audience's soul will also cause other strings to vibrate in turn. This is a stimulation of the audience's "fantasy," which "continues to create" the work.[1] These strings of the soul, which vibrate frequently, will also vibrate when other strings are sounded. And sometimes so intensely that they drown out the original sound. Some people are moved to tears when listening to "happy" music and vice versa. Therefore particular effects of a work are more or less colored by their different receptions.

But in this case the original sound is not destroyed. It lives on and continues—even if unnoticed—to work on the soul.[2]

There is no man who does not respond to art. Each work and each method of work causes in every man without exception a vibration fundamentally identical to that felt by the artist. [105]

The inner, eventually discoverable identity of the separate methods in the various arts has been the basis for an attempt to strengthen a specific sound in one art with an identical sound in another art, to reinforce it and thus to obtain an especially powerful effect. This is one way of producing effect.

But the repetition of one method of one art (e.g., music) by means of an identical method of another art (e.g., painting) is only *one* case,

[1]This "collaboration" is counted on especially in theatrical productions nowadays, although it has naturally always been planned for by the artist. This produces the desire for a certain free space between the work and its ultimate expression. This "do-not-utter-the-ultimate" was demanded by Lessing and Delacroix, among others. Space is left free for the work of fantasy.
[2]Therefore each work is correctly "understood" in the course of time.

only *one* possibility. When this possibility is also used as an inner method (e.g., in Scriabin),[3] we first consider the sphere of contrast and complicated composition as the opposite of this repetition and later as a series of possibilities that lie between effect and counter-effect. The material is inexhaustible.

The nineteenth century is distinguished as a period that lay far from inner creation. Its concentration on material appearances and on the material aspects of appearances logically caused internal creative powers to decline to the point of their virtual disappearance.

[3]See L. Sabaneiev's article in this volume.

H. Arp

This one-dimensionality naturally caused other manifestations of one-dimensionality to develop.

So it was also with the theater:

1. Here perforce (as in other fields) already existing parts (created earlier) were minutely elaborated and distinctly separated from one another, because it seemed convenient to do so. This reflects the specialization that always arises immediately when no new forms are being created, and

2. The positive character of the spirit of the time could lead only to a form of combination that was likewise positive. People thought: two is more than one, and they tried to strengthen each effect by repeating it. With inner effects this may be reversed, and often one is more than

Egyptian

two. In mathematics $1 + 1 = 2$. In the soul it is possible that $1 - 1 = 2$. [106]

Elaboration to 1. Specialization and the further external elaboration of parts is the *first consequence of materialism*. As a result three groups of stage works arose and became petrified, separated from one another by high walls: (a) drama, (b) opera, (c) ballet.

(a) Drama of the nineteenth century is, in general, a more or less subtle and profound account of events of a more or less personal character. Usually it is a description of external life; the spiritual life played a part only when it was connected with the external life.[4] *The cosmic element was entirely missing.*

*The external action and the external connection of the plot is the form of contemporary drama.*

(b) Opera is drama to which music has been added as the principal element, causing the subtlety and profundity of the dramatic aspect to suffer severely. The two parts are connected only externally. This means that either the music illustrates (or strengthens) the dramatic action, or the dramatic action helps to explain the music. [107]

This sore spot was recognized by Wagner, and he tried to improve it by various means. His basic idea was to connect the parts organically and in this way to create a monumental work.[5]

[4]There are few exceptions. Even these few (e.g., Maeterlinck, Ibsen's *Ghosts*, Andreyev's *Life of Man*, etc.) are under the spell of external events.
[5]It has taken Wagner's idea more than half a century to cross the Alps. Now it is authoritatively expressed there in the form of articles. Take the musical "manifesto"

N. Goncharova

Wagner tried to intensify the means and bring the work to a monumental height by repeating one and the same external movement in two concrete forms. His mistake was to believe that he had a universal method at his command. Actually his method is only one of a series of even more powerful possibilities of monumental art.

Parallel repetition is only *one* method, and an external repetition at that. Wagner nonetheless gave it a new form that had to lead to other forms. Before Wagner movement, for example, was entirely external and superficial in opera (perhaps only decadent). It was a naive appendage to opera: pressing the hands against the chest—love; lifting the arms—prayer; extending the arms—strong emotion, etc. Such childlike forms (which can still be seen every night) were externally connected with the libretto, which again was illustrated by the music. Wagner connected movement and musical beat directly (artistically): the movement was subordinated to the beat.

of the futurists: "We proclaim as an absolute necessity that the composer must be the author of a dramatic or tragic poem that he has to set to music" (May 1911, Milan).

195

G. Münter

This connection is still external. The inner sound of the movement plays no part. [108]

In the same artistic but still external fashion Wagner subordinated the music to the libretto, that is, to the movement in a broad sense. He represented musically the hissing of glowing iron in water, the beating of a hammer in the smithy, etc.

This *interchanged* subordination increased the methods, leading thereby to further combinations.

On the one hand Wagner increased the effect of one method, and on the other hand he decreased the inner sense, the purely artistic inner meaning, of the auxiliary method.

These forms are only mechanical reproductions (not internal parallel effects) of purposeful actions in the plot. The second connection of music and movement (in the broad sense of the word) is similar; it is the musical "characterization" of various roles. The stubborn recurrence of a motif whenever the protagonist appears on the stage is finally losing its power. It affects the ear in the same way as a well-known label on a bottle affects the eye. Feeling ultimately

P. Klee

rebels against such a consistent programmatic use of one and the same form.[6]

Finally Wagner uses the word to tell the story or to express his thoughts. However, he does not create an appropriate milieu for his purposes, because the words are usually drowned out by the orchestra. It is not enough to let the word sound in numerous recitatives. But the attempt to interrupt the incessant singing has already dealt a powerful blow to the "unity." Nevertheless, the external action remains untouched by it.

Despite Wagner's efforts to create a libretto (movement), he still completely followed the old tradition of the external, and he did not consider the third element, which is sporadically used today in a still-primitive form[7]—color and, connected with it, pictorial form (decoration). [109]

*The external action, the external connection of the various parts and of the two methods (drama and music) is the form of contemporary opera.*

(c) Ballet is a drama with all the characteristics mentioned above and with the same content. The seriousness of the drama is lost, even more so than in opera. Opera has themes other than love: religious, political, and social relationships are the grounds for enthusiasm, despair, honesty, hate, and similar feelings. Ballet contents itself with love in a childlike fairy-tale form. Besides music, single and group movements are made use of. Everything remains in a naive

[6]This programmatic use penetrates Wagner's work. Probably it may be explained not only by the artist's character, but also by his intention to discover a precise form for his new kind of creativity. The spirit of the nineteenth century gave it the "positive" stamp.
[7]See Sabaneiev's article.

197

A. Kubin

form of external connection. It is even customary for various dances to be inserted or omitted at will. The "whole" is so problematic that such practices go entirely unnoticed.

*The external action, the external connection of the various parts and of the three methods (drama, music, and dance) is the form of contemporary ballet.*

Elaboration to 2. Through the *second consequence of materialism*, i.e., positive addition $(1+1=2, 2+1=3)$, only one form of combination (or intensification) was brought about, and this demanded the same kind of methods. Thus powerful emotion was instantly underlined by a fortissimo in music. *This mathematical principle constructs the forms of effect on an entirely external basis.* [110]

All the *forms* mentioned, which I call forms of substance (drama— word, opera—sound, ballet—movement), and the combination of various methods, which I call methods for effect, were constructed to form an *external unity. All these forms originated from the principle of external necessity.*

The logical result of this is the limitation, the one-dimensionality (= impoverishment) of forms and methods. Gradually they become orthodox and each tiny change appears to be revolutionary.

198

P. P. Girieud:
*Half Nude*

Let us start on the basis of the internal. The whole state of affairs changes fundamentally.

1. Suddenly the external appearance of each element vanishes, and its inner value sounds fully.

2. Clearly, when the criterion of the inner sound is applied, the outer action obviously is not only unimportant but also creates harmful obscurity.

3. The external connection appears in its proper value, i.e., setting up unnecessary limits and weakening the inner effect.

4. Automatically the feeling of the necessity of *internal unity* is aroused. This is supported and even caused by external irregularities.

S. LUCAS EVANGEL. 3.
J800.

Bavarian Glass Painting

Egyptian

5. It opens up the possibility for each element to keep its own external life, even if it contradicts the external life of another element. [111]

If we make practical discoveries out of these abstract ones, we see that it is possible:

Elaboration to 1. To use as a method the inner sound of only one element.

Elaboration to 2. To eliminate the external action (= plot) so that

Elaboration to 3. The external connection collapses of its own accord, just like

Elaboration to 4. The external unity, and

Elaboration to 5. That the inner unity gives rise to an endless series of methods that earlier could not exist.

*The inner necessity becomes the only source.*

The following little stage composition is an attempt to draw upon this source.

There are three elements that as external methods serve the *inner value:*

1. The musical sound and its movement,

2. The physical-psychical sound and its movement, expressed through people and objects,

3. The colored tone and its movement (a special possibility for the stage).

Kandinsky: *Composition No. 5*

Van Gogh: *Portrait of Dr. Gachet*

Japanese Woodcut (Fragment)

Bavarian Glass Painting

The drama finally consists of the complex of inner experiences (soul = vibrations) of the audience.

Elaboration to 1. Music, the main element and source of the inner sound, was taken from opera. It should never be externally subordinated to the action.

Elaboration to 2. Dance was taken from ballet. It is involved with inner sound as abstract effective movement.

Elaboration to 3. The colored tone has an independent importance and is treated as a method with equal rights. [112]

All three elements play equally important roles; they remain externally independent and are treated equally, i.e., they are subordinated to the inner goal.

Music, for example, may be entirely pushed into the background or played offstage when the effect of the movement is expressive enough, and powerful musical collaboration would only weaken it. An increase of musical movement may correspond to a decrease of dance movement; in this way both movements (the positive and the negative) enhance their inner value. There are numerous combinations between these two poles: collaboration and contrast. Graphically speaking, the three movements could run in entirely separate, externally independent directions.

The word, independent or in sentences, was used to create a certain "atmosphere" that frees the soul and makes it receptive. The sound of the human voice was also used pure, i.e., without being obscured by words, or by the meaning of words.

The reader is asked to attribute the weaknesses of the following little composition, *The Yellow Sound*, not to the principle of stage compositions, but to its author's account. [113]

The Yellow Sound
A Stage Composition
*by Kandinsky*

# The Yellow Sound

## A Stage Composition[1]

## Characters:

Five Giants

Vague Creatures

Tenor (backstage)

A Child

A Man

People in Flowing Robes

People in Tights

Chorus (backstage) [117]

[1]Musical portion by Thomas v. Hartmann.

## Prelude

A few indistinct chords from the orchestra.

*Curtain*

   On the stage it is dark-blue dawn, which at first is whitish and later becomes intense dark blue. After a while, at center stage, a small light becomes visible and becomes brighter as the color deepens. After a while, orchestra music. Pause.

   Backstage a chorus is heard that must be arranged in such a way that the source of the singing cannot be located. The bass voices are to be heard above all. The singing is even, without temperament, but interrupted, as indicated in this text by dots. [119]

*First deep voices:*

"Dreams hard as stones . . . And speaking rocks . . .
Earth with riddles of fulfilling questions . . .
The motion of the heavens . . . And melting . . . of stones . . .
Invisible rampart . . . growing upward . . . "

*High voices:*

"Tears and laughter . . . Prayers while cursing . . .
The joy of union and the blackest battles."

"Dark light on the . . . sunniest . . . day

(vanishing fast and suddenly).

Blindingly bright shadow in darkest night!!"

The light vanishes. It suddenly becomes dark. A somewhat lengthy pause. Then introduction by the orchestra. [120]

French (Nineteenth Century)

German (Fifteenth Century)

# Picture 1

(Right and left of the audience.)
The stage must be as deep as possible. At rear a broad green hill.
Behind the hill a flat, mat, blue, rather deep-colored curtain.

Soon the music starts, first at a high pitch. Then suddenly and
quickly dropping lower. At the same time the backdrop turns dark
blue (simultaneously with the music) with wide black edges (like a
picture). Backstage a chorus without words becomes audible, sounding
without feeling, quite wooden and mechanical. After the chorus stops
singing, a general pause: no motion, no sound. Then darkness.

Later the same scene is lighted. From right to left five intensely
yellow giants (as large as possible) slide forth (as if gliding over the
stage).

They remain far back standing beside each other—some with
hunched shoulders, others with drooping shoulders, with strange,
indistinct, yellow faces.

They turn their heads toward each other *very* slowly and make
simple movements with their arms.

The music becomes more distinct.

Soon afterward the giants' *very* deep singing without words becomes audible (pianissimo), and the giants approach the footlights *very* slowly. Quickly from left to right fly vague red creatures, *somewhat* suggesting birds, with large heads that are remotely similar to human heads. This flight is reflected by the music. [121]

The giants continue to sing, more and more softly, becoming more and more indistinct. The hill at the rear grows slowly and becomes paler and paler. Finally white. The sky turns completely black.

Backstage the same wooden chorus becomes audible. The giants can no longer be heard.

The front of the stage becomes blue and more and more opaque.

The orchestra struggles with the chorus and defeats it.

A thick blue fog completely obscures the stage. [122]

Egyptian

Dance Mask

Russian

## Picture 2

Gradually the blue fog dissolves into an intensely white light. At the rear a completely round hill, intensely green and as large as possible.

The backdrop is purple, rather bright.

The music is shrill, stormy, with A and B and A-flat frequently repeated. The single tones are finally devoured by the noisy turbulence. Suddenly complete silence. Pause. Again A and B whimper plaintively, but distinctly and sharply. This lasts for a rather long time. Then another pause.

At this moment the backdrop suddenly turns dirty brown. The hill becomes dirty green. And exactly in the center of the hill an undefined black spot forms, which appears to be sometimes distinct, sometimes blurred. At each change the dazzling white light by jerks becomes

216

progressively grayer. On the left side of the hill a *huge* yellow flower suddenly appears. It remotely resembles a large crooked cucumber, and it becomes more and more vivid. Its stem is long and thin. Only one spiky narrow leaf grows from the middle of the stem and it is turned sideways. Long pause. [123]

Later, in *utter silence*, the flower flutters very slowly from right to left. Still later the leaf flutters also, but not with the flower. Still later they both flutter in different tempi. Then separately again. With the flower's movement a very thin B is sounded; with the leaf's, a very deep A. Then again they both flutter together, and both notes are sounded simultaneously. The flower starts trembling violently and then remains still. The orchestra continues to sound the two notes. At the same time from the left many people enter, wearing bright, long, shapeless robes (the first is entirely blue, the second red, the third green, and so on; only yellow is missing). The people have huge white flowers in their hands, similar to the one on the hill. The people keep together as closely as possible, pass close to the hill, and stop on the right side of the stage, pressed tightly against each other. They speak with mixed voices and recite:

"The flowers cover everything, cover everything, cover everything.

Shut your eyes! Shut your eyes!

We are looking. We are looking.

Cover conception with innocence.

Open your eyes! Open your eyes!

Gone. Gone."

First they speak these lines together as if in ecstasy (very distinctly). Then they repeat the same lines individually, to each other and into the distance—alto, bass, and soprano. At the line "We are looking. We are looking," a B is sounded, at the line "Gone. Gone," an A. Occasionally the voices become hoarse. Now and then one screams as if possessed. Here and there the voices become nasal, sometimes speaking slowly, at other times extremely fast. In the first case the whole stage is suddenly blanketed by a dull red light. In the second case complete darkness alternates with an intense blue light. In the third—everything suddenly becomes pale gray (all colors vanish!). Only the yellow flower glows more brightly!

Gradually the orchestra begins and overwhelms the voices. The music becomes nervous, leaps from fortissimo to pianissimo. The light becomes a little brighter, and the colors of the people can be seen vaguely. Going from right to left, tiny, imprecise figures, vaguely

Hier lieg ich als ein Kind, bis ich als Richter straff d. Sünd.

Bavarian Glass Painting

gray-green in hue, walk very slowly over the hill. They look straight ahead. When the first figure appears, the yellow flower flutters convulsively. Later it vanishes suddenly, and just as suddenly all the white flowers turn yellow.

The people move slowly to the front of the stage as if in a trance and gradually move farther away from each other. [124]

As the music fades, the same recitative is heard.[2] Soon they stop as if enraptured and turn around. They suddenly notice the tiny figures, still walking over the hill in endless succession. The people turn away and take a few quick steps to the front of the stage; they stop and turn again, and remain motionless as if chained.[3] Finally they fling away the flowers, which seem to be filled with blood, and forcibly casting off their immobility, they run to the front of the stage, close to each other. They look back frequently.[4] Suddenly it becomes dark. [125]

[2]Half a sentence is spoken in unison; at the end of a sentence *one* voice very indistinct. Changing frequently.
[3]These movements have to be executed sharply, as if on command.
[4]These movements need not be in unison.

# Picture 3

At rear: two large red-brown rocks, one pointed, the other round and larger than the first one. Backdrop: black. Between the rocks the giants (of Picture 1) are standing, whispering to each other noiselessly. Sometimes they whisper in pairs, sometimes they huddle. Their bodies remain motionless. From all sides come dazzlingly colored rays (blue, red, purple, green) alternating rapidly several times. Then all these rays become focused in the center and blend. Everything remains motionless. The giants are almost completely invisible. Suddenly all the colors vanish. The stage is black for a moment. Then a faint yellow light flows onto the stage, gradually becoming more and more intense, until the whole stage is intensely lemon yellow. As the light increases, the music becomes lower and darker (these motions suggest a snail withdrawing into its shell). While these two motions occur, nothing but light is seen on the stage, no objects. When the light is most intense, the music has faded away entirely. The giants become distinct again, but motionless, and look straight ahead. The rocks no longer are visible. Only the giants are on the stage; they are now standing farther apart from each other and have become taller. Backdrop and floor black. Long pause. Suddenly a shrill, terrified tenor voice can be heard from behind the stage, rapidly shrieking completely unintelligible words (*a* can be heard frequently, for example, Kalasimunafakola!). Pause. It becomes dark for a moment. [126]

Russian

## Picture 4

At stage left a small lopsided building (like a very simple chapel), without a door or windows. At one side of the building (on the roof) a little lopsided turret with a small cracked bell. From the bell a rope. A small boy in a white shirt sits on the floor (facing the audience) and slowly and regularly pulls the lower end of the rope. At stage right, a very large man is standing, dressed entirely in black. His face is dead white and quite indistinct. The chapel is dirty red. The turret intensely blue. The bell is made of tin. Backdrop gray, regular, flat. The black man stands with his legs apart and his arms akimbo.

The man (very loud, commanding; his voice pleasant): "Silence!!"

The child drops the rope. It becomes dark. [127]

Russian

# Picture 5

Gradually the stage is bathed in a cold red light that slowly becomes brighter and yellower. At this moment the giants in the back become visible (as in Picture 3). There are also the same rocks.

The giants are again whispering (as in Picture 3). When they again huddle together, the same shriek can be heard from backstage, but very fast and short. It becomes dark for a moment. This sequence is repeated once.[5] When it becomes light (white light, without shadow), the giants are again whispering and at the same time moving their hands weakly (these movements should each be different, but weak). Now and then one of them opens his arms (this movement should merely be suggested) and tilts his head to one side, looking at the

[5]The music must naturally be repeated each time.

Egyptian

audience. Twice the giants suddenly let their arms drop, seem to grow a little taller, and without moving look at the audience. Then a kind of convulsion seizes their bodies (as in the yellow flower), they whisper again, now and then extending their arms weakly [128] as if beseeching. Gradually the music intensifies. The giants remain motionless. From the left many people enter, clad in many-colored tights. Their hair is the same color as their tights. Their faces likewise. (The people are like puppets.) First come those in gray, then in black, in white, and finally in colors. The movements are different in each group: one walks fast, straight ahead; another, slowly as if with difficulty; a third now and then leaps joyously; a fourth looks around continually; a fifth advances in a solemn theatrical manner, arms crossed; a sixth walks on tiptoes, each with one palm raised, and so on.

They all arrange themselves differently on the stage: some sit in small secluded groups, others sit alone. In the same way some stand in groups, others alone. The whole composition of these people should neither be "beautiful" nor very definite. But it should *not* be in *complete* disorder. They look in different directions, some with heads raised, some with heads lowered, and bowed. As if oppressed by fatigue, they rarely change their positions. The light remains white. The music often changes tempo, now and then it also sags with fatigue. At that moment one of the white figures on the left (toward the back of the stage) makes vague but very rapid movements, sometimes with his arms, sometimes with his legs. At intervals he continues a motion for quite a long time and holds the pose for a few moments. It is a kind of dance. His tempo changes frequently, sometimes in unison with the music, sometimes not. (These simple movements must be very carefully worked out, so that what follows is full of expression and unexpected.) The other figures slowly begin to look at the white figure. Some crane their necks. Finally they are all looking at him. The dance ends abruptly. The white figure sits down, extends one arm as if in solemn preparation, and slowly bending his arm at the elbow, brings it to his head. The general tension becomes charged with meaning. The white figure places his elbow on his knee and rests his face on his palm. It darkens for a moment. Then the same groups and positions are seen. Some groups are lighted from above more or less brightly with different colors: one large sitting group is lighted with a bright red, a large standing group with a pale blue, and so forth. The intense yellow light is focused only on the sitting white figure (except for the giants who now become especially distinct). Suddenly all colors vanish (the giants remain yellow), and a dim white light fills the stage. In the orchestra single colors begin to speak. Corresponding to each color sound, single figures rise from different places: quickly, hastily, solemnly, slowly, and as they move, they look upward. Some remain standing. Some sit down again. After that, fatigue again overwhelms them all, and they all remain motionless. [129]

The giants whisper. But now they also remain motionless and upright, because backstage the wooden chorus becomes audible, singing for only a short time.

In the orchestra again single colors are heard. A red light sweeps over the rocks, and they tremble. Alternating with this light the giants tremble.

Movement is discernible at various points.

In the orchestra B and A are repeated several times: singly, together, sometimes very sharply, and sometimes—barely audibly.

Several figures leave their places and walk, some fast and some slowly, to other groups. Those who stood alone form smaller groups of two or three, or join larger groups. Large groups dissolve. Some figures hurry off the stage, looking backward. At the same time all black, gray, and white figures vanish: finally only the colored figures remain on the stage.

Gradually everything moves in an irregular rhythm. In the orchestra —confusion. The shrill shriek of Picture 3 becomes audible. The giants shudder. Various lights sweep the stage and cross each other.

Whole groups run offstage. A general dance starts. It begins in different places and gradually flows together, carrying all with it. Running, leaping, running to and from each other, falling. While standing, some figures rapidly move only their arms, others only their legs, or their heads, or their torsos. Some combine all these movements. *Sometimes* these are group movements. *Sometimes* whole groups make one and the same movement.

At the moment of greatest confusion in the orchestra, in movement and light, it *suddenly* becomes dark and silent. Only in the depths of the stage the yellow giants remain visible, slowly being devoured by darkness. It seems as if the giants are being snuffed out like lamps, that is, the light flickers several times before total darkness descends. [130]

Russian

# Picture 6

(This picture must appear *as quickly as possible*.)

A dull blue backdrop as in Picture 1 (without black edges).

At center stage a bright yellow giant with an indistinct white face and large, round, black eyes. Backdrop and floor black.

He slowly raises both arms (with palms facing downward) alongside his body and grows taller.

In a moment, when he extends to the full height of the stage, and his figure resembles a cross, it *suddenly* becomes dark. The music is as expressive as the action on the stage. [131]

Hall . . . dem blaßen . . . fun = det für . . . ihr

Perkussion solo

9/12. 1911

# AUS DEM „GLÜHENDEN" VON ALFRED MOMBERT

Alban Berg, Op. 2. Nº 4.

Warm die Lüf-te, es sprießt Gras auf son-ni-gen Wie - sen, Horch!

Horch es flö-tet die Nach-ti-gall.

Ich will sin - gen: Dro-ben hoch im dü-stern Bergforst, es

schmilzt und glit-zert kal-ter Schnee, ein Mäd-chen in grau-em Klei-de lehnt an feuch-tem

*) Der Vorschlag ruhig und langsam zu nehmen!          S. 9540

# „Ihr tratet zu dem herde____" aus dem „Jahr der Seele"

## von STEFAN GEORGE

### Für eine Singstimme und Klavier von ANTON von WEBERN

su - chen ta - sten ha - schen__ Wird es noch ein - mal schein!

Seht was mit trost - ge - ber - de der mond euch rät: Tre - tet

weg vom her - de, es ist wor - den spät.

# Moderne Galerie
## Heinrich Thannhauser

MÜNCHEN        THEATINERSTR. 7

Werke erster Meister
Künstler der Secessionen
Moderne Franzosen
Der blaue Reiter

# Appendix

# Documents: Facsimiles
and Translations

Sindelsdorf 10. IX. 11

Lieber Herr Piper,

beiliegendes provisorisches Inhaltsverzeichnis der 1. Nummer unseres Almanachs soll Ihnen nur berichten, daß wir in der letzten Zeit sehr fleißig an der Sache gearbeitet haben. Als neue Mitarbeiter haben wir Arnold Schönberg (der diesen Mittwoch nach München kommt u. und seine Beiträge zugesichert hat), in Paris Le Fauconnier (Vice-Präsident der indépendants), P. P. Girieud (der über die Primitiven von Siena schreibt), in Rußland als Vertreter der neuen Musik Kulbin u. Hartmann u. a. Haben Sie nicht Lust, Ihren Besuch in Murnau nun einmal zur Ausführung zu bringen? wir könnten dann gemütlich den Tag über das ganze Projekt beraten, an der Hand der zum größten Teil schon vorliegenden Beiträge. Soweit Arbeiten von lebenden Künstlern reproduziert werden, trägt jeder Künstler die Kosten seiner Klischees.

Haben Sie Kandinskys Schrift über Malerei durchstudiert? Es wäre schön, wenn er wirklich auch bald herauskäme. Im Almanach könnte man zuweilen darauf direkt Bezug nehmen, statt Gleiches zweimal zu sagen.

Kandinsky läßt Sie vielmals grüßen;

... sind es Ihnen die Überspannung der Durchschnitte
der ... zu ...? Wenn ja, so
... Ihnen ... ... ... . Wenn
... sich im ... ..., können wir
... ... nicht ... .

Und ... ... in Ihrem treu
... ... ... ...

F. Marc.

# Letter from Franz Marc to Reinhard Piper, September 10, 1911, plus "Provisional Table of Contents of the First Number"

SINDELSDORF, SEPTEMBER 10, 1911

DEAR MR. PIPER,

The enclosed provisional table of contents of the first number of our almanac is to show you that we have recently been working hard at it. As final collaborators we have Arnold Schönberg (who will come to Murnau this Wednesday and who has promised us his article); in Paris, Le Fauconnier (vice president of the Indépendants), P. P. Girieud (who is writing on the primitives in Siena); in Russia, Kulbin and Hartmann as representatives for new music. Wouldn't you like to make your visit to Murnau a reality? We would talk leisurely about the whole project all day long, on the basis of the works which are almost finished. As far as works of living artists are being reproduced, each artist will pay for his own plates.

Did you read Kandinsky's essay on painting? I wish it would soon be published. Instead of saying the same thing twice, we could refer directly to it in the almanac.

Kandinsky gives his regards. Did you send him a copy of the catalogue of the Sonderbund? If so, he thanks you very much. If you come to Murnau, we could of course talk about this too.

My regards to your wife and to yourself,

Yours,<br>F. MARC

Der blaue Reiter.

Heft I.

Probgukt.
Vorwort 1
Vorwort 2

Malerei:

1. Sienna — P. P. Girieud.
2. Konstruktion — Kandinsky.
3. Die „Wilden" Frankreichs — Le Fauconnier
4. Die „Wilden" Deutschlands — Marc.
5. Die „Wilden" Rußlands — Burljuk
6. „Neue Secession" — Hartstein.

Musik:

1. Einleitung — Kandinsky.
2. Neue Musik — Arnold Schönberg.
3. Freie Musik — N. Kulbin.
4. Farbe — Ton — Zahl — A. Unkowsky.
5. Korrespondenz aus Rußland — Th. v. Hartmann
6. Französische Musik.
7. Die neuen russischen Harmonien
8. System Jaworsky — Gottmann.

Bühne:
1. Monodrama — Jewreinoff
2. Über „Glückliche Hand" — A. Schönberg
3. Bühnenkompositionen — Kandinsky.

Chronik:
1. Ausstellung Nerses' in der alten Pinakothek München — Marc
2. Juryfragen?: München, Berlin, Wien.

Reproduktionen:
1. bayrische Glasbilder.
2. Images d'Epinal (franzö[s]. Volksblätter.
3. Russische Volksblätter.
4. Lefauconnier, Picasso, Marc, Kandinsky, Epstein, Burljuk, Münter, Bilonné, Girieud, Kokoschka, Oppenheimer, Kubin, Jawlensky, Werefkin.
5. Illustrationen um 1830.

Gegen Frühjahr 1912 erscheint bei R. Piper & Co, München:

"Der Blaue Reiter"

Die Mitarbeiter dieser in zwangloser Folge erscheinenden Organs sind hauptsächlich Künstler (Maler, Musiker, Dichter, Bildhauer)

Aus dem Inhalt des ersten Bandes:

| | |
|---|---|
| Roger Allard | Ueber die neue Malerei |
| Franz Marc | Geistige Güter |
| " " | Die "Wilden" Deutschlands |
| August Macke | Die "Masken |
| David Burljuk | Die "Wilden" Rußlands |
| Th. v. Hartmann | Die "Anarchie" in der Musik |
| Arnold Schönberg | Die Stilfrage |
| Kandinsky | "Der Gelbe Klang" (eine Bühnencomposition) |
| " | Ueber Construktion in der Malerei |
| | usw. |

Etwa 100 Reproduktionen:
Bayerische, französische, russische Volks-Kunst; primitive, römische, gotische Kunst; ägyptische Schattenfiguren, Kinderkunst usw. — Kunst des XX. Jahrhunderts: Burljuk W., Cézanne, Delaunay, Gauguin, Le Fauconnier, Girieud, Kandinsky, Kubin, Marc, Matisse, Münter, Pechstein, Picasso, Schönberg, van Gogh, Wieber etc.

Musikbeilagen. Lieder von Alban Berg,
Anton von Webern.
Preis à 10 M.
Als Herausgeber zeichnen: Kandinsky, Franz Marc.

DER
BLAUE REITER

Die große Umwälzung;
Der Verschieben der Schwerpunkten
in der Kunst, Literatur und Musik;
Die Mannigfaltigkeit der Formen:
das Constructive, Compositionelle
dieser Formen;
Die intensive Wendung zum Inneren
der Natur und der damit
verbundene Verzicht auf das
Verschönern der Äußeren der
Natur —
— das sind im Allgemeinen die
Zeichen der neuen inneren Renaissance.

x
x
Die Merkmale und Äußerungen dieser
Wendung zu zeigen,
ihren inneren Zusammenhang mit der
vergangenen Epochen hervorzuheben,
die Äußerung der inneren Bestrebungen
in jeder innerlich klingenden Form bekannt
zu machen — Ziel, welches zu erreichen
— das ist das Ziel, welches zu erreichen
„Der Blaue Reiter" sich bemühen wird.

Table of Contents, with the title "Contents of the
First Volume," from August Macke's estate,
October 1911

DER
BLAUE REITER

Inhalt der ersten Nummer.

Bitte wenden

Konstruktion in der Malerei .... Kandinsky.

<u>Reproduktionen.</u>

Ausgrabungen von Benin, Ägyptische Schattenspiele, Siamesische Schattenspiele, Negerkunst, Kinder= zeichnungen, Gotische Kunstwerke, Bayrische Glasmalereien u. Votivbilder, Russische Volksblätter und Plastik u. s. w. Moderne Kunst: W. Burljuk, R. Delaunay, LeFauconnier, Gauguin, Girieud, Cézanne, Kandinsky, Koboschka, Marc, Matisse, Münter, Jawlensky, Picasso, Werefkin u. s. w.

## Almanac: *Der Blaue Reiter* [1]

A great era has begun: the spiritual "awakening," the increasing tendency to regain "lost balance," the inevitable necessity of spiritual plantings, the unfolding of the first blossom.

We are standing at the threshold of one of the greatest epochs that mankind has ever experienced, the epoch of great spirituality.

In the nineteenth century just ended, when there appeared to be the most thoroughgoing flourishing—the "great victory"—of the material, the first "new" elements of a spiritual atmosphere were formed almost unnoticed. They will give and have given the necessary nourishment for the flourishing of the spiritual.

Art, literature, even "exact" science are in various stages of change in this "new" era; they will all be overcome by it.

Our [first and] most important aim is to reflect phenomena in the field of art that are directly connected with this change and the essential facts that shed light on these phenomena in other fields of spiritual life.

Therefore, the reader will find works in our volumes that in this respect show an *inner* relationship although they may appear unrelated on the surface. We are considering or making note not of work that has a certain established, orthodox external form (which usually is all there is), but of work that has an *inner* life connected with the great change. It is only natural that we want not death but life. The echo of a living voice is only a hollow form, which has not arisen out of a distinct *inner necessity*; in the same way, there have always been created and will increasingly be created, works of art that are nothing but

---

[1]Typescript preface by the "Editors," from August Macke's estate, October 1911. Passages in brackets were deleted in a later version. (K. L.)

hollow reverberations of works rooted in this inner necessity. They are hollow, loitering lies that pollute the spiritual air and lead wavering spirits astray. Their deception leads the spirit not to life but to death. [With all available means we want to try to unmask the hollowness of this deception. This is our second goal.]

It is only natural that in questions of art the artist is called upon to speak first. Therefore the contributors to our volumes will be primarily artists. Now they have the opportunity to say openly what previously they had to hide. We are therefore asking those artists who feel inwardly related to our goals to turn to us as *brethren*. We take the liberty of using this great word because we are convinced that in our case the establishment automatically ceases to exist.

The artist essentially works for people who are called laymen or the public and who as such have hardly any opportunity to speak. It is natural that their feelings about art and their ideas should be expressed as well. So we are ready to provide space for any serious remarks from this quarter. Even short and unsolicited contributions will be published in the "opinions" column.

[In the present situation of the arts we cannot leave the link between the artist and the public in the hands of others. Reviews are mostly sickening. Because of the growth of the daily press, many unqualified art critics have stolen in among the serious ones; with their empty words they are building a wall in front of the public instead of a bridge. We will devote one special column to this unfortunate, harmful power so that not only the artist but also the public can be enabled to see the distorted face of contemporary art criticism in a clear light.]

Works like ours do not happen at fixed intervals, nor can living creations be ordered by man. Our volumes will therefore not appear at fixed times but rather spontaneously, whenever there is enough important material.

It should be almost superfluous to emphasize specifically that in our case the principle of internationalism is the only one possible. However, in these times we must say that an individual nation is only one of the creators of all art; one alone can never be a whole. As with a personality, the national element is automatically reflected in each great work. But in the last resort this national coloration is merely incidental. The whole work, called art, knows no borders or nations, only humanity.

The Editors:
KANDINSKY, FRANZ MARC

251

Guatemalan

## Der Blaue Reiter[1]

Today art is moving in a direction of which our fathers would never even have dreamed. We stand before the new pictures as in a dream and we hear the apocalyptic horsemen in the air. There is an artistic tension all over Europe. Everywhere new artists are greeting each other; a look, a handshake is enough for them to understand each other!

We know that the basic ideas of what we feel and create today have existed before us, and we are emphasizing that in *essence* they are not new. But we must proclaim the fact that everywhere in Europe new forces are sprouting like a beautiful unexpected seed, and we must point out all the places where new things are originating.

Out of the awareness of this secret connection of all new artistic production, we developed the idea of the *Blaue Reiter*. It will be the call that summons all artists of the new era and rouses the laymen to hear. The volumes of the *Blaue Reiter* are written and edited exclusively by artists. The first volume herewith announced, which will be followed at irregular intervals by others, includes the latest movements in French, German, and Russian painting. It reveals subtle connections with Gothic and primitive art, with Africa and the vast Orient, with the highly expressive, spontaneous folk and children's art, and especially with the most recent musical movements in Europe and the new ideas for the theater of our time.

[1]Text of the subscription prospectus, written by Franz Marc in mid-January 1912.

Cover Design for the *Blaue Reiter* by Wassily Kandinsky, water color, 11″ x 8″ (Nina Kandinsky, Neuilly-sur-Seine).

Cover Design for the *Blaue Reiter* by Wassily Kandinsky, water color, $10\frac{3}{4}''$ x $8\frac{2}{3}''$ (Städtische Galerie, Munich).

Cover Design for the *Blaue Reiter* by Wassily Kandinsky, water color, 10¾″ x 8⅔″ (Städtische Galerie, Munich).

Henri Rouſſeau.

# DER BLAUE REITER

Mit etwa 140 Reproduktionen. Vier handkolor. graph. Blätter.

Bayeriſche, ruſſiſche Volkskunſt; primitive, römiſche, gotiſche Kunſt; ägyptiſche Schattenfiguren, Kinderkunſt. — Kunſt des XX. Jahrhunderts: Burljuck W., Cézanne, Delaunay, Gauguin, Girieud, Kandinsky, Kokoſchka, Kubin, Le Fauconnier, Marc, Matiſſe, Münter, Picaſſo, Henri Rouſſeau, Schönberg, van Gogh uſw.

MUSIKBEILAGEN: Lieder von Alban Berg, Arnold Schönberg. A. von Webern. — Herausgeber: Kandinsky und Franz Marc.

Es erſcheinen drei Ausgaben: Allgemeine Ausgabe: geh. M 10.—, geb. M 14.—. Luxus-Ausgabe: 50 Exemplare. Enthält noch zwei von den Künſtlern ſelbſt kolorierte und handſignierte Holzſchnitte. Preis M 30.—. Muſeums-Ausgabe: 10 Exemplare: jedem Exemplar wird eine Originalarbeit eines der beteiligten Künſtler beigegeben. Preis M 100.—. Nur vom Verlag direkt zu beziehen. Proſpekte koſtenlos.

Advertisement by the publishing house, early March 1912(?).

# Preface to the Second Edition

Two years have passed since this book first appeared. One of our aims
—to me the main one—has remained virtually unattained. It was to
demonstrate through examples, practical juxtapositions, and theoretical proofs that the question of form in art was secondary, that
the question of art was primarily one of content.

In practice the *Blaue Reiter* was right: its formal creation is dead.
It has lived—or ostensibly lived—for scarcely two years. But the
necessity for its existence has "developed" further. Thanks to the
rashness of our time, the more easily understandable ideas have
formed "schools." So the movement represented here has generally
become broader and yet more compact as well. The explosions
necessary for the initial breakthrough are abating in favor of a quieter,
increasingly stronger, broader, more compact current.

This spreading of the spiritual movement, as well as its strong concentric force that powerfully attracts more and more new elements,
manifests its natural destiny and its visible goal.

Thus life, reality, goes its own way. The thunderous characteristics
of a great era are almost inexplicably ignored; the public (including
many art theoreticians), in opposition to the spiritual trend of the
time, more than ever continues to consider, to analyze, to systemat-
ize, the formal element exclusively. Maybe the time has not yet come
for "hearing" and "seeing."

But the justified hope that that time will come is rooted in necessity.

And this hope is the most important reason for a further edition
of the *Blaue Reiter*.

In the course of these two years we have come closer to the future
in particular instances. Precision and evaluation have become even
more possible. Everything else grows organically out of the general

idea. This development, this particularly clear relationship of spiritual fields that were formerly distinctly separate, their mutual approach, and their occasional mutual penetration resulting in mixed and therefore richer forms—all these factors demonstrate the necessity for a further development of these ideas in a new publication.

K[ANDINSKY]

"Everything that comes into existence on earth can be only a beginning."

This statement by Däubler could well be the motto for our work and our intentions. There will be a fulfillment sometime, in a new world, in another existence. On earth we can furnish only the theme. This first book is the opening note of a new theme. The alert listener must have sensed the meaning of the book in its disconnected, restlessly moving manner. He found himself near some headwaters where, at a hundred different places at once, there was mysterious bubbling, now hidden, now openly singing and murmuring. With a divining rod we searched through the art of the past and the present. We showed only what was alive, what was not touched by the tone of convention. We gave our ardent devotion to everything in art that was born out of itself, lived in itself, did not walk on crutches of habit. We pointed to each crack in the crust of convention only because we hoped to find there an underlying force that would one day come to light. Some of these cracks have closed again, our hope was in vain; out of others a lively spring is now gushing. But this is not the only reason for the book. It has always been the great consolation of history that nature continuously thrusts up new forces through outlived rubbish. If we saw our task as simply pointing out the natural spring of a new generation, we could calmly leave this to the course of time; there would be no need to conjure up the spirit of a great epoch of change with our cries.

We say No to the great centuries. We know that with this simple denial we cannot stop the serious methodical development of the sciences and triumphant "progress." Also we do not even dream of anticipating this development, but to the scornful amazement of our contemporaries, we take a side road, one that hardly seems to be a road, and we say: this is the main road of mankind's development.

We know that the great mass cannot follow us today; the path is too steep and too far from the beaten track for them. But a few already do want to walk with us. The fate of this first book taught us this. Now we let the book go forth again unchanged, while we ourselves are detached from it and involved in new projects. We do not know yet when we will get together for the second book. Perhaps only when we are entirely alone again, when the cult of modernity has stopped trying to industrialize the virgin forest of new ideas. Before the second book is completed, many things that fastened onto the movement in those years must be cast off or torn away, even by force. We know that everything could be destroyed if the beginnings of a spiritual discipline are not protected from the greed and dishonesty of the masses. We are struggling for pure ideas, for a world in which pure ideas can be thought and proclaimed without becoming impure. Only then will we or others who are more talented be able to show the other face of the Janus head, which today is still hidden and turns its gaze away from the times.

We admire the disciples of early Christianity who found the strength for inner stillness amid the roaring noise of their time. For this stillness we pray and strive every hour.

*March 1914*                                   F[RANZ] M[ARC]

# Foreword to the Planned Second Volume of the *Blaue Reiter*

Once more and *many times more* we are trying to divert the attention of ardent men from the nice and pretty illusion inherited from the olden days toward existence, horrible and resounding.

Whenever the leaders of the crowds turn right, we turn left; when they point to a goal, we turn our backs; whatever they warn us against we hurry toward.

The world is crammed to suffocation. On every stone man has put the brand of his cleverness. Every word is leased or invested. What can man do for salvation but give up everything and flee? What but draw a dividing line between yesterday and today?

This is the great task of our time—the only one worth living and dying for. Not the slightest contempt for the great past is involved in this. We want something else. We do not want to live as carefree heirs or to live on the past. Even if we wanted to live like that, we could not. The inheritance is used up, and substitutes are making the world base.

Therefore we venture forth into new fields, and we are shocked to find that everything is still untrodden, unspoken, uncultivated, unexplored. The world lies virginal before us; our steps are shaky. If we dare to walk, we must cut the umbilical cord that ties us to our maternal past.

The world is giving birth to a new time; there is only one question: has the time now come to separate ourselves from the old world? Are we ready for the *vita nuova*? This is the terrifying question of our age. It is the question that will dominate this book. Everything in this volume is related to this question and to nothing else. By it alone should we measure its form and its value.

*February 1914*                                          Fz. Marc

# Notes on the Contributors

ROGER-CHARLES-FÉLIX ALLARD, b. 1885 in Paris; poet, writer, and art critic. His contribution to the *Blaue Reiter*, "Signs of Renewal in Painting," was translated by Franz Marc.

1. Ref.: Hector Talvart and Joseph Place, *Bibliographie des auteurs modernes de langue française (1801–1927)* (Paris, 1928), I, 58–61. (See further references there.)

DAVID BURLIUK, b. 1882 in Kharkov, d. 1967 in the United States; painter and writer, brother of Vladimir Burliuk. Attended art academies in St. Petersburg and Odessa; in 1903 he studied with Azbé in Munich, in 1904–05 with Cormon in Paris. In 1907 he returned to Russia and became one of the leaders of the artistic avant-garde. He was co-founder of the groups The Blue Rose (1907) and The Donkey's Tail (1911–12).

A member of the Neue Künstlervereinigung München, he, together with his brother, wrote a text for its second exhibition. Left Russia in 1918.

2. Refs.: Katherine S. Dreier, *Burliuk*, foreword by Duncan Phillips (New York, 1944). Reproductions selected by Marcel Duchamp and Katherine S. Dreier.
*Color and Rhyme*, No. 31 (1956).
Camilla Gray, *The Great Experiment: Russian Art 1863-1922* (London and New York, 1962), pp. 69ff.

ERWIN RITTER VON BUSSE, b. 1885; son of an officer. After attending a military college, he entered the army as a cadet. On passing the officer's examination, he was promoted to lieutenant in 1906. In the following year he received his discharge and went to Munich to study law and later history of art. In the Bavarian capital he became acquainted with modern trends in painting. He was profoundly influenced by Wilhelm Worringer's essay *Abstraktion und Einfühlung (Abstraction and Empathy)*, published in 1908. In fall 1912 he enrolled at the University of Bern, from which he received his Ph.D. early in 1914. His letters to Robert Delaunay from the period of the *Blaue Reiter* are part of the painter's estate.

3. Ref.: Erwin Ritter von Busse, *Entwicklungsgeschichte des Problems der Massendarstellungen in der italienischen Malerei* (Ph.D. dissertation, University of Bern [Munich, 1914]). Author's biography on p. 76.

EUGÈNE DELACROIX, b. 1798, d. 1863; French painter. The context for the quotation from Delacroix's *Journal* of 1857 is as follows: "Most writing on art is by people who are not artists: thus all the misconceptions. I believe that everybody who has received a good education can speak adequately about a book, but not about a painting or a piece of sculpture." Quoted from Eugène Delacroix, *Mein Tagebuch*, Bibliothek ausgewählter Kunstschriftsteller No. 2 (Berlin, 1903), p. 200. This edition was in the possession of Franz Marc. In 1936 Kandinsky wrote in *Cahiers d'Art*: ". . . with their utterances Delacroix and Goethe support our ideas. . . ."

JOHANN WOLFGANG VON GOETHE, b. 1749, d. 1832; German poet, dramatist, novelist. The quotation is from a conversation with Riemer on May 19, 1807. From *Goethe im Gespräch*, ed. Franz Deibel and Friedrich Gundelfinger (Friedrich Gundolf) (3d enlarged ed. [Leipzig, 1907], p. 94). Cf. also Kandinsky's remarks: "These supposed rules, which will soon lead to a 'thorough bass' in painting, are merely the recognition of the inner effect of various methods and their combination" ("On the Question of Form," p. [90]).
". . . the deep relations among the arts, and especially between music and painting. Goethe said that painting must consider this relation its ground [thorough bass] and by this prophetic remark he foretold the position of painting today" (*Concerning the Spiritual in Art* [New York: George Wittenborn, 1947], p. 46).
"Because of the very nature of modern structure, there has never been a time when it was more difficult than it is today to formulate a complete theory or to construct a pictorial foundation [thorough bass]. . . . However, it would be rash to say that there are no principles in painting comparable to a foundation, or that such principles would inevitably lead to academicism" (Ibid., p. 67).

THOMAS VON HARTMANN, b. 1883 or 1886 in Moscow, d. 1956 in New York; composer, pianist, and painter. After graduating from the Moscow Conservatory, he lived in Munich from 1908 to 1911; in 1922 he settled in Paris. From 1951 to his death he lived in New York. Hartmann wrote several works for orchestra, piano, and voice, music for the stage and the ballet, two operas, and chamber music. He not only applied new methods in composition but, like his painter friends in the Neue Künstlervereinigung München and the *Blaue Reiter*, struggled for a synthesis of all artistic genres (e.g., music for the dances of Alexander Sacharov). In an (unpublished) speech delivered in New York in 1950, he gave an account of his friendship with Kandinsky. For the contribution "On Anarchy in Music," see pp. [43ff.].

262

4. Refs.: *Prospekt der Werke Hartmanns* (New York [1950?]).
   *The Macmillan Encyclopedia of Music and Musicians* (New York, 1938), p. 771.
   *Grove's Dictionary of Music and Musicians* (New York, 1955), IV, 125.
   Will Grohmann, *Wassily Kandinsky: Life and Work* (New York, 1958), passim.

EUGEN KAHLER, b. 1882 in Prague, d. 1911 in the same city; painter graphic artist, and poet. From 1901 to 1905 he studied with Heinrich Knirr and Franz Stuck at the Academy of Art in Munich. He also took private lessons with Hugo Habermann. In 1906 and 1907 he lived in Paris. In 1910 he returned to Munich, where his works were exhibited at Thannhauser's gallery. Kandinsky's obituary in the almanac is the best eulogy of his work and of his short life. Kandinsky also included two large prints in the First Exhibition of the Editors of the *Blaue Reiter* (Nos. 22 and 23).

WASSILY KANDINSKY, b. 1866 in Moscow, d. 1944 in Neuilly-sur-Seine; according to his self-characterization, "painter, graphic artist, and writer— the first painter to base painting on purely pictorial means of expression and also to banish subject matter from pictures" (*Das Kunstblatt* [1919], p. 172). Kandinsky turned to painting only after studying law and political economy and teaching at the University of Dorpat. He came to Munich in 1896, attended Azbé's art school, then the Academy, where he studied with Franz Stuck. In 1901 he became president of the Phalanx artists' group. From 1908 on, after several trips—to France, Tunis, Italy—he alternately lived in Munich and Murnau on Lake Staffel. In 1909 he became co-founder of the Neue Künstlervereinigung München. Cf. our text about this period and that of the *Blaue Reiter*. In 1914 Kandinsky returned to his native country via Switzerland and Sweden. After the Revolution he held several leading administrative positions in art; in 1921 he emigrated to Berlin. He worked at the Bauhaus in Weimar and Dessau for more than ten years. Late in 1933 he moved to France.

5. Refs.: [Wassily] Kandinsky, *Über das Geistige in der Kunst* (*Concerning the Spiritual in Art*) (Munich, 1912). Also several later editions and translations.
   [Wassily] Kandinsky, *Punkt und Linie zu Fläche* (*Point and Line to Plane*), Bauhaus Book No. 9 (Munich, 1926). Also several later editions and translations.
   Will Grohmann, *Wassily Kandinsky: Life and Work* (New York, 1958). Basic monograph with catalogue of work and bibliography.
   Kunsthalle, Basel, *Wassily Kandinsky 1866–1944. Gesamtausstellung* (1963). The exhibition was also shown in New York, Paris, and The Hague.
   Jacques Lassaigne, *Kandinsky*, Le goût de notre temps (Geneva, 1964).

N. KULBIN, dates of his life unknown; physician and professor at the St. Petersburg Medical Military Academy with the title of general. Also graphic artist and patron of avant-garde artists. In 1910 he published an article entitled "Studio of the Impressionists." Friend of David Burliuk.

6. Refs.: *Color and Rhyme*, No. 31 (1956), p. 20.

Katherine S. Dreier, *Burliuk*, foreword by Duncan Phillips (New York, 1944), pp. 52ff., 61. Reproductions selected by Marcel Duchamp and Katherine S. Dreier.

Camilla Gray, *The Great Experiment: Russian Art 1863–1922* (London and New York, 1962), pp. 98, 288.

Dimitri Chitsevski, interviews.

MIKHAIL KUZMIN, b. 1876 in Yaroslavl, d. 1936; poet, graphic artist, and composer. Kuzmin was an opponent of symbolism and founded "clarism." He wrote various prose works, and published short books of poetry. A few small volumes of short stories were translated into German. Besides musical comedies, he composed music for Grillparzer's *Die Ahnfrau*. After 1930 he was no longer mentioned in Russian literary criticism.

7. Refs.: *Brockhaus*, II (1924), 745.

Arthur Luther, *Geschichte der russischen Literatur* (Leipzig, 1924), pp. 428ff.

Dimitri Chitsevski, interviews.

AUGUST MACKE, b. 1887 in Meschede, southern Westphalia, d. 1914 (killed in action); painter. Grew up in Bonn. After attending the Düsseldorf Academy and Arts and Crafts School, he painted scenery for the theater in Düsseldorf. On a trip to Paris in 1907 he found himself. Bernhard Koehler, uncle of his future wife, had made this trip possible. After studying briefly with Lovis Corinth in Berlin, he returned to Paris the following year. After his marriage, in 1909, he lived in Bonn; in January 1910 he met Franz Marc, with whom he formed a close friendship. Marc introduced him to the Munich *Blaue Reiter* group. In 1912, during his third trip to Paris, he and Marc visited Robert Delaunay. Shortly before the outbreak of the war, he went on his famous journey to Tunis, together with Paul Klee and Louis Moilliet. The result of this trip was a series of enchanting water colors.

In his eulogy to the artist on his untimely death, Franz Marc wrote: "Compared to us he gave color the brightest and purest sound, as clear and bright as his whole being." For the essay "Masks," see pp. [21ff.].

8. Refs.: Gustav Vriesen, *August Macke* (Stuttgart, 1953; 2d ed., 1957).

Elisabeth Erdmann-Macke, *Erinnerung an August Macke* (Stuttgart, 1962).

August Macke and Franz Marc, *Briefwechsel*, ed. Wolfgang Macke, DuMont Dokumente (Cologne, 1964).

FRANZ MARC, b. 1880 in Munich, d. 1916 (killed in action) at Verdun; **Notes**
painter and writer. After abandoning his plans to study theology, he
enrolled in 1900 in the Art Academy of Munich, where he studied with
Wilhelm von Diez and G. Hackl. In 1903 he traveled to Paris and Brittany,
in 1906 to Athos and Salonika. At Easter of 1907 he returned to Paris,
where he was strongly impressed by the paintings of Van Gogh. From 1909
he made his home in Sindelsdorf, in Upper Bavaria. Early the following
year he met August Macke, and the two became close friends. In Munich
he became associated with the Neue Künstlervereinigung München (cf.
pp. 11ff. for the art controversies and incidents that led to the founding of the
Editors of the *Blaue Reiter* together with Kandinsky). On his third trip to
Paris—together with Macke—in October 1912, Marc met Robert Delaunay
whose theory of "simultaneity" he adapted for his own late works. In 1913
he played an important role in Herwarth Walden's Erster deutscher Herbstsalon (First German Fall Salon). In spring 1914 he purchased a home at
Ried near Benediktbeuern. On March 4, 1916, he was killed on the Western
Front. Franz Marc is one of the noblest figures of German art. His paintings
enjoy exceptional popularity; his notes and letters from the front are valued
as treasures of twentieth-century German literature.

9. Refs.: Franz Marc, *Briefe, Aufzeichnungen und Aphorismen* (Berlin, 1920).
   2 vols.

   Alois J. Schardt, *Franz Marc* (Berlin, 1936).

   Hermann Bünemann, *Franz Marc: Zeichnungen und Aquarelle* (Munich,
   1948; 3d ed., 1960).

   Klaus Lankheit, *Franz Marc*, ed. Maria Marc (Berlin, 1950).

   Idem, "Franz Marc—Der Mensch und das Werk," in *Unteilbares Sein:
   Aquarelle und Zeichnungen von Franz Marc* (Cologne, 1959), pp. 12ff.

   Idem, *Franz Marc: Der Turm der blauen Pferde*, Reclam Werkmonographie
   No. 69 (Stuttgart, 1961).

   Kunstverein, Hamburg, *Franz Marc: Ausstellung im Kunstverein in Hamburg
   1963–64* (1963). Catalogue with bibliography of Marc's published
   writings.

WASSILY WASSILIEVICH ROZANOV, b. 1859, d. 1919, well-known writer, author
of numerous books primarily on philosophy of religion strongly tinged
with sexology. Decisively influenced by Dostoevsky.

10. Refs.: Arthur Luther, *Geschichte der russischen Literatur* (Leipzig, 1924),
    p. 421.

    V. V. Zenkovsky, *History of Russian Philosophy* (Paris, 1948), I, 457ff.
    In Russian; also in French and English (New York, 1953).

    Wassily Rozanov, *Solitaria*, selected writings, trans. with introductory
    essay by Heinrich Stammler (Hamburg, 1963). In German.

    Dimitri Chitsevski, interviews.

LEONID LEONIDOVICH SABANEIEV, b. 1881 in Moscow; composer and writer on music. Sabaneiev studied with Taneyev; stylistically he is a disciple of Moussorgsky. In 1921 he became a member of the board of directors of the State Institute of Musicology, which was founded at his suggestion; in 1924 he emigrated to France; later he lived in England and America. Among other works, he composed two piano trios, one violin sonata, one piano sonata, and several pieces for piano. His main writings include a work on Scriabin (Moscow, 1916) and a history of Russian music (Moscow, 1924); abroad he published *Modern Russian Composers* (New York, 1927) and a book on his teacher, Taneyev (Paris, 1930). Kandinsky wrote to Marc on December 31, 1911: ". . . Sabaneiev's article on Scriabin is most interesting and will surely be impressive. Hartmann and I worked conscientiously at the translation all last night. I hope to finish it today."

11. Refs.: Hugo Riemann, *Musiklexikon* (1961), II, 560.

O. Thompson, ed., *The International Cyclopedia of Music and Musicians* (New York, 1946), p. 1860.

ARNOLD SCHÖNBERG, b. 1874 in Vienna, d. 1951 in Los Angeles; one of the great masters of "new music," largely self-taught. From 1901 to 1903 he lived in Berlin, teaching at the Stern Conservatory. Later he moved to Vienna, where he both conducted and taught. There, in 1911, his main theoretical work, *Structural Functions of Harmony*, was published. After World War I, in which he served as an Austrian soldier for some time, he moved to Mödling near Vienna. In 1925, after he had discovered the twelve-tone technique, he was appointed to teach advanced composition at the Prussian Academy of the Arts in Berlin. In 1933 he lost his position, was forced to leave the country, and emigrated to the United States, where he taught music at the University of California from 1936 to 1944. He had started painting as early as 1908. Kandinsky valued his paintings so highly that he wrote an article on them and included three studies in oil in the First Exhibition of the Editors of the *Blaue Reiter*. Cf. illustrations, pp. [80] and [86].

12. Refs.: *Arnold Schönberg* (Munich, 1912). Contributions by Alban Berg, Paris von Gütersloh, K. Horwitz, Heinrich Jalowetz, Wassily Kandinsky, Paul Königer, Karl Linke, Robert Neumann, Erwin Stein, Anton von Webern, Egon Wellesz.

H. H. Stuckenschmidt, *A. Schönberg* (Zurich and Freiburg, 1951; 2d ed., 1957).

J. Rufer, *Das Werk Arnold Schönbergs* (Kassel-Basel-London-New York, 1959).

Florence. *Dipinti e Disegni di Arnold Schönberg*. (Opened May 27, 1964.) Exhibition catalogue with an Italian translation of Kandinsky's essay on Schönberg's paintings from the 1912 volume of tributes.

# Notes on the Illustrations*

## A. Reproduced in Color in Original Edition

1. Bavarian Mirror Painting [frontispiece]                                      50
   St. Martin and the Beggar; redrawing of a mirror painting; hand colored.
   Original formerly owned by Kandinsky or Münter, now in Munich, Städtische Galerie im Lenbachhaus. Reproduced in *Vasily Kandinsky: Painting
   on Glass* (New York: The Solomon R. Guggenheim Museum, 1966), p. 9,
   fig. 2.

   Unnumbered: Wassily Kandinsky, Woodcut [preceding 1]                          53
   Hand colored and signed by the artist; included only in the de luxe edition,
   not listed in table of contents.

   Unnumbered: Franz Marc, Woodcut [preceding 1]                                 54
   Hand colored and signed by the artist; included only in the de luxe
   edition, not listed in table of contents.

   Unnumbered: Franz Marc, *Horses* [preceding 33]                              100
   Only in the de luxe edition and the second edition. After a water color
   now in Hamburg, Kunsthalle. Also in Marc's estate, proof colored by the
   artist.

2. Franz Marc, *Horses* [preceding 33]                                         101
   After a water color in R. Piper's archives. In the second edition replaced
   by a second version, now in Hamburg Kunsthalle. Draft sketch in Hanover,
   Collection Sprengel (exhibition catalogue 1965, no. 188).

3. Wassily Kandinsky, Study for *Composition No. 4* [preceding 65]             139
   Probably after a water color now owned by Mr. and Mrs. Emerson
   Woelffler, Colorado Springs, Colorado.
   Ref.: Will Grohmann, *Wassily Kandinsky: Life and Work* (New York, 1958),
   p. 406, ill. No. 683 (1911).

4. Figure from Egyptian Shadow Play [preceding 97]                             179
   Hand colored; not from *Der Islam* (cf. n. for No. 23) but from an original
   owned and made available by Kahle.

---

*This annotated listing replaces the register in the original almanac (which
began on [page 135]).

267

# B. Reproduced in Black and White in Original Edition

5. German Woodcut [1]    55

From *Ritter vom Turn, von den Exempeln der Gottesforcht und erbarkeit* (The Knight of the Tower; Examples of the Fear of God and of Honesty) (Basel: Michael Furter, 1495), as reproduced in Worringer, ill. No. 52.

6. Chinese Painting [2]    56

Probably not a Chinese painting but a nineteenth-century imitation; original unknown.

7. Bavarian Mirror Painting [4]    57

Death of a Saint; Raymundsreut, Bavarian Forest; after 1800; mirror painting 10″ x 6½″. Oberammergau, Heimatmuseum der Gemeinde (formerly Collection Krötz/Murnau—Collection Kapfer/Murnau). Photo: Robert Braunmüller, Munich.

The Krötz/Kapfer Collections were purchased by the Museum of the Municipality of Oberammergau after World War II. Nine originals for the reproductions in the almanac were discovered.

8. Pablo Picasso, *Femme à la guitare* [following 4]    58

Paris, spring 1914; oil on canvas 21½″ x 14¾″ (22¼″ x 16½″). Prague, Collection Dr. Vincenc Kramar. Photo: Galerie Louise Leiris, Paris.

Ref.: Christian Zervos, *Pablo Picasso*, II, Part 1 (Paris, 1942), No. 237.

9–10. Two Drawings by Children [preceding 5]    59

The origin of the drawings by children and amateurs (ills. 9, 10, 36, 51, 78-81, 99-102) cannot be ascertained.

11. August Macke, *Storm* [5]    61

Sindelsdorf; oil on canvas 33½″ x 44½″; signed "A. Macke 1911" and on the back "Aug. Macke 1911." Saarbrücken, Saarland-Museum. Photo: Staatliche Bildstelle für das Saarland, Saarbrücken.

Refs.: Catalogue of the First Exhibition of The Editors of the *Blaue Reiter*, No. 27 (ill.); G. Vriesen, *August Macke* (2d ed.; Stuttgart, 1957), No. 290.

12. Ernst Ludwig Kirchner, *Women Dancers* [6]    62

Four women dancers, 1910; lithograph 12½″ x 16″; signed bottom right: "E. L. Kirchner." Bremen, Collection Wolfgang Budzcies. Photo: Stickelmann, Bremen.

Ref.: Catalogue of the Second Exhibition of the Editors of the *Blaue Reiter*, Nos. 86–96 (ill.); Gustav Schiefler, *Die Graphik Ernst Ludwig Kirchners* (Berlin-Charlottenburg, 1926), I, No. 167.

13. Ironwood Sculpture from South Borneo [7]    63

Dayak. Figure of ancestor carved and polished from red-brown rosewood (trunk). With feather headgear, shield, and spear. This figure is placed in front of house and village entrances as a guardian against evil demons; height 76″. Bern, Bernisches Historisches Museum, Ethnographische Abteilung. Photo: Bernisches Historisches Museum, Bern.

14. Illustration from Grimm's *Fairy Tales* [8]	66	**Notes**

Franz Marc's information is incorrect (cf. also text p. [10]). There is no 1832 edition of Grimm's *Fairy Tales*. Nor can this illustration be found in any other edition. Cf. *Anmerkungen zu den Kinder- und Hausmärchen der Brüder Grimm*, newly revised by Johannes Bolte and Georg Polivka (Leipzig, 1930), IV, 473ff. (Information kindly supplied by Arthur Henkel.) The title "Reinhald das Wunderkind" clearly points to the Musäus tradition. But the corresponding illustration in Musäus' folk fairy tales—first published in 1842 and drawn by Adolf Schröther—is not identical with our picture. It was probably taken from an almanac that included the Grimm fairy tale "The Three Sisters." (Information kindly supplied by Kurt Ranke.) For those almanac illustrations, cf. Maria Gräfin Lanckorónska and Arthur Rümann, *Geschichte der deutschen Taschenbücher und Almanache aus der klassisch-romantischen Zeit* (Munich, 1954).

15. Wassily Kandinsky, *Lyrical* [9]	67

1911; oil on canvas 37″ x 51″; signed bottom right: "Kandinsky." Rotterdam, Museum Boymans-van Beuningen. Photo: Museum Boymans-van Beuningen.

Ref.: Will Grohmann, *Wassily Kandinsky: Life and Work* (New York, 1958), p. 353, ill. No. 51 (1911).

16. Heinrich Campendonk, *Jumping Horse* [10]	68

1911; oil on canvas 33½″ x 25½″. Saarbrücken, Saarland-Museum. Photo: Reichmann, Saarbrücken.

Ref.: Catalogue of the First Exhibition of the Editors of the *Blaue Reiter*, No. 14 (ill.).

17. Bavarian Mirror Painting [11]	69

Birth of Christ, Raymundsreut, Bavarian Forest, after 1800; mirror painting 10″ x 6¼″. Oberammergau, Heimatmuseum der Gemeinde (formerly Collection Krötz/Murnau—Collection Kapfer/Murnau). Photo: Robert Braunmüller, Munich.

18. Bavarian Glass Painting [12]	70

*Winter*, from a series of the seasons; Seehausen, Upper Bavaria, first half of the nineteenth century; glass painting 9½″ x 6½″. Oberammergau, Heimatmuseum der Gemeinde (formerly Collection Krötz/Murnau—Collection Kapfer/Murnau). Photo: Robert Braunmüller, Munich. The caption at the bottom, *Winter*, was omitted in the reproduction.

19. Mosaic [following 12]	71

Apparition of St. Mark's Body; Venetian-Byzantine style. Venice, Cathedral of San Marco, right aisle, thirteenth century. Photo: Alinari, Florence.

20. Japanese Drawing [13]	72

From *Kôrin Gashiki* (published 1818), a book of woodcuts after pictures by Kôrin (1658–1716).

21. Russian Folk Sculpture [14]                                                  73
1911; formerly in the possession of Kandinsky(?).
22. Vladimir Burliuk, *Portrait Study* [15]                                      74
About 1911; oil on canvas, size unknown; according to David Burliuk,
lost 1917 in Russia.
Ref.: Catalogue of the First Exhibition of the Editors of the *Blaue Reiter*,
No. 12 (ill.).
23. Figure from Egyptian Shadow Play [16]                                        75
*Der Islam*, II (1911), 175, fig. 70.
Kahle's essay appeared in: *Der Islam*, I (1910), 264–99; II (1911), 143–95;
also cf. Paul Kahle, *Zur Geschichte des arabischen Schattentheaters in
Ägypten* (Leipzig, 1909).
24. Russian Folk Print [following 16]                                            76
Eight prints exhibited in the Second Exhibition of the Editors of the
*Blaue Reiter*, Nos. 258–65 (1 ill.).
Ref.: For these prints in general: Irina Danilova, *Der russische Volks-
bilderbogen* (Dresden, 1962).
25. Russian Folk Print [following 17]. Cf. n. for No. 24.                        77
26. David Burliuk, *Head* [18]                                                   78
Portrait of the artist's mother, about 1910; oil on canvas, size unknown
(second version in private American collection; *Der Blaue Reiter*, exhibition
catalogue [Leonard Hutton Galleries, New York, 1963], No. 41). According
to the artist, the painting was left in Russia in 1917 and has since been lost.
Ref.: Catalogue of the First Exhibition of the Editors of the *Blaue Reiter*,
No. 9 (ill.).
27. Japanese Drawing [19]                                                        79
Probably European copy. Original unknown.
28. German Lithograph [20]                                                       81
Original unknown. Probably original for a glass painting of a St. George.
Lithograph 14″ x 10″, by F. Wentzel, Wissembourg, ca. 1840–50.
29. Head of a Stonemason [following 20]                                          82
Magdeburg Cathedral, about 1210; detail from a column capital in the
chancel aisle (portrait?). Photo: Dr. F. Stoedtner–H. Klemm, Düsseldorf.
Refs.: W. Greischel, *Der Magdeburger Dom* (Berlin, 1929), pp. 31 and 60;
K. Gerstenberg, *Die deutschen Baumeisterbildnisse des Mittelalters* (Berlin,
1966). According to F. Arens, *Das Münster*, XX (1963), 329, probably a
prophet.
30. Jocuno [21]                                                                  83
Dance mask, representing the head of a tapir; woven reed covered with
bark fiber, base color white, with black, brown, and yellow overpainting;
South America, Yuri-Taboca Indians, height 20″. Munich, Staatliches
Museum für Völkerkunde, Inv. No. 372 (Spix and Martius). Photo: Staatliches Museum für Völkerkunde, Munich.

Ref.: *Indianer vom Amazonas*, Exhibition of the Staatliches Museum für **Notes**
Völkerkunde (Munich, 1960), catalogue No. 92, col. ill. No. 9.

31. Sculpture from Easter Islands [22]                                      84
Figure representing ancestor; wood (probably Toomiro wood), height 12″.
Munich, Staatliches Museum für Völkerkunde, Inv. No. 193. Photo:
Staatliches Museum für Völkerkunde, Munich.

32. Sculpture from Cameroons [23]                                          86
Board carved on both sides, 68½″ x 13″; the significance of the carving is
unknown. Munich, Staatliches Museum für Völkerkunde, Inv. No. 93.13.
Photo: Staatliches Museum für Völkerkunde, Munich.

33. Sculpture from Mexico [24]                                             87
Figure of the god Xipe Totec, "The Flayed God." In his left hand a round
shield, in his right hand the rattle wand with magic powers. Made in a mold,
the back is flat. Huexotla, Aztec. Yellow-gray clay. Height 6¼″. In original
edition, the illustration was reversed. Munich, Staatliches Museum für
Völkerkunde, Inv. No. 10.1713. Photo: Staatliches Museum für Völkerkunde, Munich.
Ref.: *Alt-Amerika. Präkolumbische Kunst aus Mexiko, den Maya-Ländern,
dem südlichen Mittelamerika, Kolumbien und Ecuador*, catalogue for the
exhibition of the Staatliches Museum für Völkerkunde, ed. Andreas
Lommel, rev. Otto Zerries (Munich, 1964), No. 36.

34. Sculpture from New Caledonia [25]                                      87
Wooden mask, height 24½″. Munich, Staatliches Museum für Völker-
kunde, Inv. No. 02.230. Photo: Staatliches Museum für Völkerkunde,
Munich.

35. Chieftain's Cape from Alaska [26]                                      88
Poncho of mountain-goat wool, woven in a characteristic manner and
attached to a leather backing cut into fringes at the bottom. Black-yellow-
white pattern with the typical eye ornaments of the Northwestern American
Indians, probably the Tlingit (Chilcat) tribe. Length 36¼″. Munich, Staatliches Museum für Völkerkunde, Collection Leuchtenberg, No. 779. Photo:
Staatliches Museum für Völkerkunde, Munich.

36. *Arabs*, Drawing by a Child [following 26]                            89
Cf. n. for Nos. 9–10.

37. German Book Illustration [27]                                         90
From *Eunuch des Terenz* (The Eunuch by Terence) by Konrad Dinkmuth
(Ulm, 1486); reproduced in Worringer, ill. No. 26.

38. Ivory Sculpture [28]                                                  91
Vanity (Death and Maiden), France about 1450; ivory, height 5¾″.
Munich, Bayerisches Nationalmuseum.
Refs.: Rudolf Berliner, *Die Bildwerke in Elfenbein*, Catalogues of the
Bayerisches Nationalmuseum, XIII, 4 (Munich, 1926), No. 81, plate 35

(listed as: German [Rhine?] about 1530–40); *Kunst und Kunsthandwerk. Meisterwerke im Bayerischen Nationalmuseum München*, Festschrift for the one-hundredth anniversary of the Museum (Munich, 1955), No. 28, ill. 28.

39. Alfred Kubin, Ink Drawing [29]     92
Whereabouts unknown.

40. Bavarian Glass Painting [30]     93
Beheading of a priest; Seehausen, Upper Bavaria, first half of nineteenth century; painting on glass 9″ x 6¼″. Oberammergau, Heimatmuseum der Gemeinde (formerly Collection Krötz/Murnau—Collection Kapfer/Murnau). Photo: Robert Braunmüller, Munich.

41. Figure from Egyptian Shadow Play [31]     95
*Der Islam*, II (1911), 173, fig. 68.

42. Wooden Sculpture [32]     96
Stilt; Marquesas Islands, height 12¼″. Munich, Staatliches Museum für Völkerkunde, Inv. No. 188. Photo: Staatliches Museum für Völkerkunde, Munich.

43. Bavarian Glass Painting [following 32]     97
Mary with the Son of God. Whereabouts unknown. Probably not from Collection Krötz, since it is the only illustration for August Macke's manuscript of which a plate previously existed (in August Macke's estate).

44. Robert Delaunay, *The Eiffel Tower* [following 32]     98
1911; oil on canvas, size unknown. Formerly Berlin, Collection Koehler (destroyed in World War II).
Ref.: Catalogue of the First Exhibition of the Editors of the *Blaue Reiter*, No. 17 (ill.).

45. El Greco, *St. John* [preceding 33]     99
Oil on canvas 43¼″ x 27½″. Formerly Berlin, Collection Koehler (destroyed in World War II).
Refs.: *Katalog der Sammlung Koehler* (1925), No. 150; Harold E. Wethey, *El Greco and His School*, II: catalogue raisonné (Princeton, 1962), No. X-379 ("School or workshop of El Greco, ca. 1600–1610").

46–47. Malayan Painted Wooden Figures [34]     103
Statuettes, representing husband and wife, both with crown and sarong, the husband holding a piece of fruit in his hand; Bali, carved wood painted in many colors; heights: husband 9″, wife 8¾″. Bern, Bernisches Historisches Museum, Ethnographische Abteilung. Photo: Bernisches Historisches Museum, Bern.

48. Paul Cézanne, from *Seasons* [35]     104
*Fall*, 1859/62; oil on canvas 123¾″ x 41″. Paris, Ambroise Vollard. Photo: Réunion des Musées Nationaux, Château de Versailles.
Ref.: Lionello Venturi, *Cézanne—son art, son oeuvre* (Paris, 1936), No. 6.

49. Henri Le Fauconnier, *The Swamp* [37]     106
Formerly Collection Poliakov, Moscow. Present whereabouts unknown.

50. Henri Matisse, *La Danse* [38]                                    107 **Notes**
1910; oil 102½″ x 153½″. Moscow, Museum of Western Art (from Collection Shukin). Photo: APN, Moscow.
Refs.: Alfred H. Barr, *Matisse: His Art and His Public* (New York, 1951), p. 362; Gaston Diehl, *Henri Matisse* (Paris, 1954), plate 48.

51. Drawings by Children [39]                                         108
"Arranged by adults for a frieze." Cf. n. for Nos. 9–10.

52. Paul Cézanne, from *Seasons* [40]                                 110
*Winter*, 1859/62; oil on canvas 123¾″ x 41″. Paris, Ambroise Vollard. Photo: Réunion des Musées Nationaux, Château de Versailles.
Ref.: Lionello Venturi, *Cézanne—son art, son oeuvre* (Paris, 1936), No. 7.

53. Beheading of St. John [41]                                        111
Silk embroidery from Lower Saxony, fourteenth century; formerly Kunstgewerbemuseum, Berlin. According to information from the directors of the Kunstgewerbemuseum, Berlin—Staatliche Museen/Stiftung Preussischer Kulturbesitz, the embroidery was lost in World War II.

[53a.] Unidentified Saint                                            112
Not mentioned in original edition. Woodcut, 1818. See Introduction, n. 21.

54. Malayan Painted Wooden Sculpture [43]                            113
Statuette; from their costumes and jewelry, the mother and child depicted are of princely rank; Bali; carved wood, painted in many colors, height 22″. Bern, Bernisches Historisches Museum, Ethnographische Abteilung. Photo: Bernisches Historisches Museum, Bern.

55. August Macke, Ballet Sketch [44]                                 114
About 1910; brush drawing; size and whereabouts unknown.
Ref.: Catalogue of the Second Exhibition of the Editors of the *Blaue Reiter*, Nos. 131–42 (ill.).

56. Hans Baldung Grien, Woodcut [44]                                 115
Stallions fighting in the midst of a group of wild horses in the forest, woodcut 9″ x 13½″; signed and dated on the little plate at the bottom right: "Baldung 1534." Interpreted together with two similar prints as a portrayal of "Passion and Bitterness of Love of All Creatures." Photo: Staatliche Kunsthalle, Karlsruhe.
Ref.: *Hans Baldung Grien*, exhibition catalogue (Karlsruhe, 1959), No. II H, 80.

57. Albert Bloch, *Impression of Sollnhofen* [45]                    116
About 1911. Technique and size unknown; according to the artist's widow, whereabouts unknown.

58. Erich Heckel, *Circus* [46]                                      117
1910; lithograph 10½″ x 13″; signed bottom right: "E. Heckel 1910." Hamburg, Collection Schiefler. Photo: Kleinhempel, Hamburg.
Ref.: Catalogue of the Second Exhibition of the Editors of the *Blaue Reiter*, Nos. 40–42 (ill.).

274

Ref.: Catalogue of the First Exhibition of the Editors of the *Blaue Reiter*,
No. 11.

69. Paul Gauguin, Wood Relief [60]                                        132
   *Pape Moe* (*Mysterious Water*); oak, carved and painted, 32″ x 24¼″.
Collections: Gustave Fayet, Béziers; Antoine Fayet, Védilhan; Gray's
information "Bührle Zürich" is incorrect. Photo: Réunion des Musées
Nationaux, Château de Versailles.

The theme is obviously a scene from Gauguin's diary *Noa-Noa*: "Tout à
coup, à un détour brusque, j'aperçus dressée contre la paroi du rocher
qu'elle caressait plutôt qu'elle ne s'y retenait de ses deux mains, une jeune
fille, nue: elle buvait à une source jaillissante, très haute dans les pierres . . ."
(*Noa-Noa*, facs. edition, p. 87).

Ref.: Christopher Gray, *Sculpture and Ceramics of Paul Gauguin* (Baltimore,
1963), No. 107.

70. Classical Relief [preceding 61]                                        133
   Gorgon as mistress of the animals; bronze 16¾″ x 23¼″; Etruscan Archaic,
from the Perugia region. Munich, Glyptothek, Antikensammlungen.
Photo: Antikensammlungen, Munich.

Ref.: *Kunst und Leben der Etrusker*, exhibition catalogue (Cologne, 1956),
No. 253.

71. Emil Nolde, Stage Sketch [64]                                          136
   Theater (small), Berlin 1910; water color, size unknown. According to
Stiftung Emil und Ada Nolde, whereabouts unknown.

Ref.: Catalogue of the Second Exhibition of the Editors of the *Blaue Reiter*,
Nos. 211–14 (ill.).

72. Japanese Drawing [68]                                                  140
   Obviously more Japanese in form than 27 and 105. Origin questionable.

73. Max Pechstein, *Bathers* [70]                                         142
   *Bathers III*, 1911; color woodcut 13″ x 15¾″; hand-colored impressions
in red, green, yellow, and blue; 15 impressions on Japan paper, exhibited by
Fritz Gurlitt, Berlin.

Refs.: Catalogue of the Second Exhibition of the Editors of the *Blaue Reiter*,
No. 215 (ill.); Paul Fechter, *Das graphische Werk Max Pechsteins* (Berlin,
1921), No. 54.

74. Henri Le Fauconnier, *Abundance* [following 70]                       143
   Nov./Dec. 1910; oil on canvas 75″ x 48½″; signed bottom left: "Le
Fauconnier." The Hague, Gemeentemuseum. Photo: Foto Marburg,
Berlin.

75. Wilhelm Morgner, Drawing [72]                                         144
   Brickyard; large size, exact measurements unknown. Whereabouts
unknown.

Ref.: Catalogue of the Second Exhibition of the Editors of the *Blaue Reiter*,
No. 157 or 158 (ill.).

90. Bavarian Votive Painting (Church in Murnau) [82]  162
91. Bavarian Votive Painting (Church in Murnau) [82]  163
For Nos. 88–91, cf. n. for No. 77.
92. Henri Rousseau, *Malakov: Telegraph Poles* [83]  164
1908; oil on canvas 18″ x 21¾″; signed bottom right: "Henri Rousseau 1908." Bern, private collection. Photo: André Rosselet, Auvernier-Neuchâtel, Switzerland.
Ref.: Jean Bouret, *Henri Rousseau* (Munich, 1963), No. 199.
93. Arnold Schönberg, *Self-Portrait* [85]  166
1911; oil on cardboard 19¼″ x 17″. Los Angeles, Collection Mrs. Gertrud Schönberg. Photo: Galerie Maeght, Paris.
Ref.: Catalogue of the First Exhibition of the Editors of the *Blaue Reiter*, No. 41.
94. Unknown Master [following 86]  167
Spanish? Formerly Collection Koehler (destroyed in World War II).
95. Henri Rousseau, *View of Fortifications* [87]  169
1909; oil on canvas 18″ x 21¾″; signed bottom left: "Henri Rousseau." Private collection. Photo: A. Rosselet, Auvernier-Neuchâtel, Switzerland.
Ref.: Jean Bouret, *Henri Rousseau* (Munich, 1963), No. 208.
96. Figure from Egyptian Shadow Play [89]  170
*Der Islam*, II (1911), p. 164, fig. 53.
97. Franz Marc, *The Bull* [following 90]  171
1911; oil on canvas 39¾″ x 53″; signed and dated on the back: "Fz. Marc 11." New York, The Solomon R. Guggenheim Museum (formerly Berlin, Collection Koehler). Photo: The Solomon R. Guggenheim Museum, New York.
Ref.: Alois J. Schardt, *Franz Marc* (Berlin, 1936), No. I/1911/23.
98. Japanese Wash Drawing [91]  172
Probably a sketch ascribed to Hokusai (1760–1849), but the origin has not been ascertained.
99–102. Four Heads, Drawings by Amateurs [92-93]  173
Cf. n. for Nos. 9–10.
103. Henri Rousseau, *Rousseau with Lamp* [94]  174
About 1899; oil on canvas 9¼″ x 7½″. France, Collection Pablo Picasso. Photo: André Rosselet, Auvernier-Neuchâtel, Switzerland.
Ref.: Jean Bouret, *Henri Rousseau* (Munich, 1963), No. 118.
104. Henri Rousseau, *The Wedding* [95]  175
1905; oil on canvas 64″ x 44¾″; signed bottom right: "Henri Julien Rousseau." Paris, Collection Mme. Jean Walter. Photo: André Rosselet, Auvernier-Neuchâtel, Switzerland.
Ref.: Jean Bouret, *Henri Rousseau* (Munich, 1963), No. 27.
105. Japanese Ink Drawing [96]  176
Apparently a European copy. Original unknown.

106. Alfred Kubin, *The Fisherman* [following 96]          177
Done probably at the turn of 1911/12; ink drawing on old assessor's paper 8¾" x 6" (drawing), 12¼" x 7" (sheet size); signed on bottom right margin: "Kubin"; titled on bottom left margin: "Der Fischer." Marked in pencil on the back by Kandinsky: "Bl. R. Strichmanier! Sehr vorsichtig behandeln!" (*"Blaue Reiter* line technique! Handle with utmost care!"). Karlsruhe, private collection.

107. Gabriele Münter, *Still Life with St. George* [98]          181
1911; oil on cardboard 20" x 26¾"; signed bottom left: "Münter." Munich, Städtische Galerie im Lenbachhaus. Photo: Galerie Maeght, Paris.

108. Japanese Ink Drawing [100]          182
Apparently a European copy. Original unknown.

109. *Foolish Virgin* [following 100]          183
Sandstone, from Paradise Door of Magdeburg Cathedral, c. 1240–50.
Ref.: W. Greischel, *Der Magdeburger Dom* (Berlin, 1929), pp. 55 and 61.

110. Oskar Kokoschka, *Portrait of Else Kupfer with Dog* [following 100]          184
About 1910; oil on canvas 35½" x 28"; signed top right: "OK." Zurich, Kunsthaus. Photo: Kunsthaus, Zurich.
Ref.: H. M. Wingler, *Oskar Kokoschka* (Salzburg, 1956), No. 49 and pl. 14.

111. Henri Rousseau, *Portrait of the Artist's Wife* [preceding 101]          185
Only in first edition. About 1899; oil on canvas 8¾" x 6¾". France, Collection Pablo Picasso. Photo: Verlag DuMont Schauberg, Cologne.
Ref.: Jean Bouret, *Henri Rousseau* (Munich, 1963), No. 119.

112. Bavarian Mirror Painting [101]          188
Stigmatization of St. Francis of Assisi; Raymundsreut, Bavarian Forest, after 1800; mirror painting 10" x 6¼". Oberammergau, Heimatmuseum der Gemeinde (formerly Collection Krötz/Murnau—Collection Kapfer/Murnau). Photo: Robert Braunmüller, Munich.

113. Folk Painting [102]          189
St. Anthony of Padua, eighteenth/nineteenth century; votive picture on wood, size unknown. Formerly Murnau, Catholic Parish Church of St. Nikolaus. Stolen from the church after World War II, together with ten other votive pictures (information from the Parish Office).

114. Chinese(?) Mask [103]          190
Pongwe mask from Gabon; light wood, height 13¼", width 8¼". Bern, Bernisches Historisches Museum/Ethnographische Abteilung. Photo: Bernisches Historisches Museum, Bern.

115. Wooden Sculpture from the Lower Rhine (?) [104]          192
Formerly Berlin, Collection Koehler. Destroyed in World War II.

116. Hans Arp, *Sketch for a Bust* [105]          193
1911/12; brush drawing, size unknown; whereabouts unknown.

Ref.: Catalogue of the Second Exhibition of the Editors of the *Blaue Reiter*, No. 3 (ill.).

117. Figure from Egyptian Shadow Play [106]     194
*Der Islam*, II (1911), 167, fig. 57.

118. Natalia Goncharova, *Grape Harvest* [107]     195
About 1911/12; pencil drawing 11¼" x 14½". Paris, Collection Mme. Nina Kandinsky. Photo: Galerie Maeght, Paris.
Ref.: Catalogue of the Second Exhibition of the Editors of the *Blaue Reiter*, No. 27 (ill.).

119. Gabriele Münter, *Man at the Table* [108]     196
1911; oil on cardboard 20¼" x 27"; signed bottom right: "Münter." Munich, Städtische Galerie im Lenbachhaus. Photo: Städtische Galerie, Munich.

120. Paul Klee, *Stonecutter* [109]     197
1910 (no. 74); ink and wash; size unknown. According to Felix Klee, whereabouts unknown.
Ref.: Catalogue of the Second Exhibition of the Editors of the *Blaue Reiter*, No. 106 (ill).

121. Alfred Kubin, Ink Drawing [110]     198
Whereabouts unknown.

122. Pierre Paul Girieud, *Half Nude* [following 110]     199
Formerly Berlin, Collection Koehler. Destroyed in World War II.

123. Bavarian Glass Painting [111]     200
St. Luke (from a series of the Evangelists), Upper Bavaria, dated 1800; painting on glass 12¼" x 8". The legend at the bottom: "S. Lucas Evangel. 3. 1800," was deleted for the reproduction in the almanac. Oberammergau, Heimatmuseum der Gemeinde (formerly Collection Krötz/Murnau—Collection Kapfer/Murnau). Photo: Robert Braunmüller, Munich.

124. Figure from Egyptian Shadow Play [112]     201
*Der Islam*, II (1911), 176, fig. 73.

125. Wassily Kandinsky, *Composition No. 5* [following 112]     202
1911; oil on canvas 74¾" x 108¼"; signed bottom left: "Kandinsky 1911." Solothurn, Collection Müller. Photo: R. Spreng, Basel.
Refs.: Catalogue of the First Exhibition of the Editors of the *Blaue Reiter*, No. 24; Will Grohmann, *Wassily Kandinsky: Life and Work* (New York, 1958). Catalogue of Works, No. 144 (1911).

126. Vincent van Gogh, *Portrait of Dr. Gachet* [112]     203
Painted in Auvers, June 1890; oil on canvas 26¾" x 22½". Paris, Louvre; another version from the same period, Faille No. 753, now New York. Collection Kramarsky. Photo: Réunion des Musées Nationaux, Château de Versailles.
Ref.: J. B. de la Faille, *L'Oeuvre de Vincent van Gogh* (Paris and Brussels, 1938), II, No. 754.

## Vignettes

The caricature of the German publishers on page 24 is by an unknown
artist. The drawing is in the Reinhard Piper archives, Munich.

The design for the announcement on page 238 was drawn by Franz Marc.

# Notes on the Musical Compositions

1. "The Fond Heart" (text by Maurice Maeterlinck) for soprano, celeste, harmonium, and harp, by Arnold Schönberg     226
   Cf. notes to Schönberg's article in text.
2. From "The Ardent Lover" (text by Alfred Mombert) by Alban Berg  234
   ALBAN BERG, b. 1885 in Vienna, d. 1935 in Vienna; composer. With Anton von Webern the most important of Arnold Schönberg's disciples and one of the major representatives of musical expressionism. His best-known work is the opera *Wozzeck* after Büchner's dramatic fragment.
   Literature: Willi Reich, *Alban Berg*, including writings by Berg and contributions by Theodor Adorno and Ernst Krenek (Vienna, 1937); H. F. Redlich, *Alban Berg* (Vienna-Zurich-London, 1957).
3. "You Reached the Hearth . . ." (text by Stefan George) by Anton von Webern     236
   ANTON VON WEBERN, b. 1883 in Vienna, d. 1945 in Mittersill; composer. Disciple of Arnold Schönberg and also the composer whose work in serial music has for decades been the focus of interest for the younger generation.
   Literature: Anton Webern, *Dokumente—Bekenntnisse* (Die Reihe. Informationen über serielle Musik, No. 2) (Vienna, 1955); Anton Webern, *Wege zur neuen Musik*, ed. Willi Reich (Vienna, 1960); Walter Kolneder, *Anton Webern* (=Kontrapunkte 5) (Rodenkirchen, 1961); Riemann, *Musiklexikon* (1961), II, 898ff.

# Selected Bibliography

*by Bernard Karpel*

Librarian, The Museum of Modern Art

Material grouped as *General Works* (nos. 13–40).—*Articles and Reviews*
(nos. 41–58).—*Exhibition Catalogues* (nos. 59–73).—*Individual References*
(nos. 74–94). (For nos. 1–12, see Lankheit references, pp. 261–66.)


## General Works

13. *Der Blaue Reiter*. Editors: Kandinsky, Franz Marc. München, R. Piper
& Co., 1912. (29.5 x 23 cm)
  140 pp., 142 illus. (4 hand colored), 8 vignettes and initials, 3 musical
pieces, colored cover design by Kandinsky. *Second edition, 1914:* Forewords
by the editors; 1 color plate added, 1 black and white omitted. Three editions advertised: "Allgemeine Ausgabe".—"Luxus-Ausgabe" (50 copies).—
"Museums-Ausgabe" (10 copies). See announcement, p. 256. Reprint, bibl.
24.
14. Brion, Marcel. *German Painting*. New York, Universe Books, 1960;
Paris, Pierre Tisné [1959].
  Section on the *Blaue Reiter*, Macke, Marc, Klee, and Kandinsky.
15. Buchheim, Lothar-Günther. *Der Blaue Reiter und die "Neue Künstlervereinigung München."* Feldafing, Buchheim, 1959.
  With chapters on Marc, Kandinsky, etc., and chronology on the *Blaue
Reiter*. Includes contemporary photos, overall documentation, pp. 313–40.
16. Buchheim, Lothar-Günther. *The Graphic Art of German Expressionism*.
New York, Universe Books, 1960.
  Section on the *Blaue Reiter*, pp. 46–51; bibliography. Original edition:
*Graphik des deutschen Expressionismus*. Feldafing, Buchheim, 1959. Docu-
mentation, pp. 267–90.
17. Chipp, Herschel B., ed. *Theories of Modern Art*. A Source Book by
Artists and Critics. Contributions by Peter Selz and Joshua C. Taylor.
Berkeley and Los Angeles, University of California Press, 1968.
  Kandinsky's "On the problem of form" (pp. 155–70) is translated by
Kenneth Lindsay from *Der Blaue Reiter* (pp. 74–100). Other references
passim.

18. *Dictionary of Modern Painting.* General Editors: Carlton Lake and Robert Maillard. 3 ed. New York, Tudor [1964?].

Contributions by various specialists, e.g., "Blaue Reiter" by Jean Leymarie, pp. 31–33; "Marc" by Franz Meyer, pp. 219–20, etc. First edition: Hazan, Paris [1954?]; 1st English ed. New York, Tudor [1955?].

19. Eichner, Johannes. *Kandinsky und Gabriele Münter: von Ursprungen moderner Kunst.* Munich, Bruckmann, 1957.

Chapter on the *Blaue Reiter.*

20. Grote, Ludwig. *Deutsche Kunst im zwanzigsten Jahrhundert.* 2d ed. Munich, Prestel, 1954.

"Der Weg zum Blauen Reiter," pp. 33–40. ill. First edition issued for exhibition at Lucerne, 1953.

21. Hamilton, George Heard. *Painting and Sculpture in Europe, 1880–1940.* Baltimore, Md., Penguin Books, 1967.

Index, p. 425. Bibliographies.

22. Haftmann, Werner. *Painting in the Twentieth Century.* New York, Praeger, 1965. 2 vol.

Paperback edition. Original edition: *Malerei im 20. Jahrhundert.* Munich, Prestel, 1954–55. Vol. 1: Text.—2: Pictorial supplement. Biographies, bibliography. American translation: New York, Praeger, 1960. 2 vol.

23. Händler, Gerhard. *German Painting in Our Time.* Berlin, Rembrandt, 1956.

Section on the *Blaue Reiter*; biographies. Translated from the German.

24. Lankheit, Klaus, ed. *Der Blaue Reiter, herausgegeben von Wassily Kandinsky und Franz Marc. Mit 160 Abbildungen. Dokumentarische Neuausgabe von Klaus Lankheit.* Munich, R. Piper & Co., 1965.

Piper paperback (364 pp., no color), cf. bibl. 13. Original text of the 1912 edition, pp. 13–249. Also forewords to 1914 edition. English edition by Thames and Hudson, London, 1974; The Viking Press, New York, 1974, with supplemental bibliography.

25. Lindsay, Kenneth. *An Examination of the Fundamental Theories of Wassily Kandinsky.* Doctoral dissertation. Madison, Wis., University of Wisconsin, 1951.

Typescript. Bibliography. Appendixes include translation: Program of the Society of Painters, St. Petersburg, Dec. 10, 1912.

26. Miesel, Victor H., ed. *Voices of German Expressionism.* Englewood Cliffs, N.J., Prentice-Hall, 1970.

"Der Blaue Reiter," pp. 43–88. Bibliographical notes.

27. Munich, Städtische Galerie. *Der Blaue Reiter in der Städtische Galerie im Lenbachhaus München.* Munich, The Gallery, 1963. 256 pp., ill.

Permanent collection of Kandinsky and other associated artists. Brief biographies. Excerpts from texts, letters, critiques. Includes essay from H. K. Roethel: *Moderne deutsche Malerei* (translated in bibl. 32).

28. Myers, Bernard S. *The German Expressionists: a Generation in Revolt.* New York, McGraw-Hill, 1963.

Concise edition. Original edition with extensive bibliography, New York, Praeger, 1957. Also continental editions. Paperback: New York, Praeger, 1966.

29. Neigemont, Olga. *German Expressionists: the Blue Rider School.* New York, Crown, 1968.

Brief introduction; 24 illus. incl. col.; biographies. Translated from the German; copyright Uffici Press, Lugano (1968).

30. Raynal, Maurice, ed. *The History of Modern Painting*, Vol. 2: *Matisse, Munch, Rouault, Fauvism, Expressionism.*—Vol. 3: *From Picasso to Surrealism.* Geneva, Skira, 1949–50.

Vol. 2 covers Kandinsky and has associated bibliographies; for the German edition a special expressionist bibliography was compiled by Hans Bolliger. Vol. 3 covers Kandinsky, Marc, Klee, the *Blaue Reiter*, etc.; associated bibliographies.

31. Ritchie, Andrew Carnduff, ed. *German Art of the Twentieth Century*, by Werner Haftmann, Alfred Hentzen, William S. Lieberman. New York, Museum of Modern Art, 1957.

General article on painting by Werner Haftmann includes section on "The Blaue Reiter," pp. 54–70. General and special bibliographies, pp. 227ff. Issued on the occasion of the exhibition, winter 1957.

32. Roethel, Hans Konrad. *Modern German Painting.* New York, Reynal, [1957].

Artists' statements; bibliography. Translated by Desmond and Louise Clayton from the German. Also Wiesbaden, Vollmer, 1958.

Extract in bibl. 27.

33. *The Selective Eye.* Edited by Georges and Rosamund Bernier. New York, Reynal, 1956.

Includes Will Grohmann: The Blue Rider, pp. 26–35, translation of bibl. 43.

34. Selz, Peter. *German Expressionist Painting.* Berkeley and Los Angeles, University of California Press, 1957.

Chapters on the *Blaue Reiter*, Kandinsky, etc. General bibliography, pp. 354–70.

35. Sihare, Laxmi P. *Oriental Influences on Wassily Kandinsky and Piet Mondrian, 1909–1917, a Dissertation.* New York, Institute of Fine Arts, New York University, 1967. 2 vol.

Typescript. On the accessibility of Oriental doctrines and publications, both from original sources and European theosophy, and their relevance to Kandinsky's thought, writing, and works. Extensive documentation and bibliography.

36. Tietze, Hans. *Lebendige Kunstwissenschaft. Zur Krise der Kunst und der*

*Kunstgeschichte.* Vienna, Krystall, 1925.

"Der Blaue Reiter" (from *Kunst,* 1919). Also note bibl. 56.

37. Vollmer, Hans. *Allgemeines Lexikon der bildenden Künstler des XX. Jahrhunderts.* Leipzig, Seemann, 1953–62.

Compact biographies and bibliographies cover the *Blaue Reiter* artists. Supplements U. Thieme and F. Becker. *Allgemeines Lexikon der bildenden Künstler.* Leipzig, Seemann, 1907–47.

38. Volpi-Orlandi, M. (Marisa Volpi Orlandini?). *Kandinsky e il Blaue Reiter.* Milan, 1970.

Preceded by: *Kandinsky dall'Art Nouveau alla psicologia della forma.* Rome, 1968.

39. Walden, Nell, and Schreyer, Lothar. *Der Sturm: Ein Erinnerungsbuch an Herwarth Walden und die Künstler aus dem Sturmkreis.* Baden-Baden, Woldemar Klein, 1954.

Index to *Der Sturm,* pp. 211ff.—List of publications and exhibitions: "Erste Ausstellung, März 1912, Der Blaue Reiter."

40. Wingler, Hans Maria. *Der Blaue Reiter.* Feldafing, Buchheim, 1954.

"Zeichnungen und Graphik von Marc, Kandinsky, Klee, Macke, Jawlensky, Campendonk, Kubin." Also variant second edition with Münter, color plates, etc.

# Articles and Reviews

*L'Arte Moderna.* See bibl. 57.

41. Le "Cavalier Bleu." *Derrière le Miroir* (Paris) nos. 133–134. Oct.–Nov. 1962.

Special double number of de luxe bulletin of the Galerie Maeght. Articles by W. Grohmann, P. Volboudt, and H. H. Stuckenschmidt. Chronology.

42. Gassiot-Talabot, Gérald. The Blaue Reiter situation. *Cimaise* (Paris) v. 9, no. 62, pp. 12–23 incl. ill. Nov.–Dec. 1962.

Text in French, English, German, and Spanish.

43. Grohmann, Will. Le Cavalier Bleu. *L'Oeil* (Paris) no. 9, pp. 4–13, ill. (col., port.). Sept. 1955.

Translated for *The Selective Eye* (bibl. 33).

44. Grote, Ludwig. Der Blaue Reiter. *Die Kunst* (Munich) v. 48, no. 1, pp. 4–11, ill. 1950.

Also note bibl. 20.

45. Grote, Ludwig. Les peintres du Blaue Reiter. *Art d'Aujourd'hui* (Paris) v. 4, no. 6, pp. 2–3, ill. Aug. 1953.

46. Kandinsky, Wassily. Der Blaue Reiter: Rückblick. *Das Kunstblatt* (Berlin) v. 14, pp. 57–60. 1930.

Partly translated in bibl. 72.

47. Kurdibowsky. [Lecture based on Kandinsky's art theories, especially *Der Blaue Reiter*] In: Program, December 10, Part I. St. Petersburg, Society of Painters, 1912.
Program translated in Lindsay dissertation, bibl. 25.

48. Lankheit, Klaus. Zur Geschichte des Blauen Reiters. *Der Cicerone* (Leipzig) v. 3, pp. 110–14. 1949.
Supplemented by: Bibel-Illustrationen des Blauen Reiters. *Anzeigen des Germanischen Nationalmuseums* (Nuremberg), 1963, pp. 199ff.

49. Lindsay, Kenneth. The genesis and meaning of the cover design for the first *Blaue Reiter* exhibition catalogue. *The Art Bulletin* (New York) v. 35, no. 1, pp. 47–52, ill. Mar. 1953.

50. Nishida, Hideho. Kandinsky: genèse du Cavalier Bleu. *XXe Siècle* (Paris) no. 28, pp. 18–24 ill. (col.). Dec. 1966.

51. Podestà, Attilio. Il "Cavaliere azzurro." *Emporium* (Bergamo) v. 112, pp. 74–78, ill. Aug. 1950.

52. Ringbom, Sixten. Art in "the epoch of the great spiritual": occult elements in the early theory of abstract painting. *Journal of the Warburg and Courtauld Institutes* (London) v. 29, pp. 386–418. 1966.
Also note Sihare, a contemporary study founded on European and Oriental analysis (bibl. 35).

53. Roditi, Edouard. Interview with Gabriele Münter. *Arts* (New York) v. 34, pp. 36–41. Jan. 1960.
Includes comment on the *Blaue Reiter* at the Tate Gallery.

54. Roters, Eberhard. Vasily Kandinsky und Die Gestalt des Blauen Reiters. *Jahrbuch der Berliner Museum* v. 5, pp. 201ff. 1963.

55. Thwaites, John A. The Blaue Reiter, a milestone in Europe. *The Art Quarterly* (Detroit) v. 13, no. 1, pp. 12–20. Winter 1950.

56. Tietze, Hans. Der Blaue Reiter. *Die Kunst für Alle* (Munich) v. 27, pp. 543ff. 1911–12.
Also note bibl. 36.

57. Veronesi, Giulia. Wassily Kandinsky, Franz Marc, Il Cavaliere Azzurro. *L'Arte Moderna* (Milan) v. 6, n. 47; pp. 41–80. 1967.
Special number; numerous illustrations (col. port.). Additional material in other issues: v. 6, n. 54, pp. 323–36, illus. (Kandinsky and the Blue Rider). —v. 14, n. 129, p. 555 (index to series passim [under "B", not "C"]). No reference to bibliography, p. 606–607, or biographies, v. 14, nos. 127–128–129.

58. Volboudt, Pierre. La cinquantenaire du Blaue Reiter. *XXe Siècle* (Paris) no. 18, suppl. pp. 6–8; 24, 26, 28, 30, 32. Feb. 1962.
Includes English summary. Also preface to his brochure: *Kandinsky, 1896-1921.* Paris, Hazan; New York, Tudor, 1963.

Bibliog-
raphy

287

# Exhibition Catalogues

59. Basel. Kunsthalle. *Der Blaue Reiter, 1908–14: Wegbereiter und Zeitgenossen.* Jan. 21–Feb. 26, 1950.

"Der Weg zum Blauen Reiter" (pp. 2–5).—"Der Blaue Reiter" (pp. 5–13) (Ludwig Grote).—"Tagebuchnotizen von Kandinsky, Marc und Klee," pp. 13–15.

60. Berlin. Der Sturm. *Erste Ausstellung: Der Blaue Reiter.* Mar. 1912.

See note under bibl. 69. Galerie Der Sturm also included the *Blaue Reiter* in its "Erster deutscher Herbstsalon," Sept.–Nov. 1913 and organized a circulating exhibition in 1914 (bibl. 63).

61. Berne. Gutekunst and Klipstein. *Ausstellung von der "Brücke" zum "Bauhaus."* Jan. 21–Feb. 18, 1956.

Graphics and documents include the *Blaue Reiter.*

62. Edinburgh. Royal Scottish Academy. *The Blue Rider Group.* An exhibition organized with the Edinburgh Festival Society by the Arts Council of Great Britain. Aug. 20–Sept. 18, 1960.

Introduction by H. K. Roethel, pp. 5–12.—Documents, pp. 23–28.—Biographical and bibliographical notes on the artists, p. 29–31. Also shown at the Tate Gallery, Sept. 30–Oct. 30. For London reviews see E. Hoffmann, *Burlington Magazine*, Nov. 1960, pp. 498–501.—D. Duerden, *Art News and Review*, Oct. 8–22, 1960, pp. 1, 12.

63. Helsingfors. Der Sturm. *Der Blaue Reiter.* Feb.–Mar. 1914.

Also: Trondheim (Apr.–May 1914).—Göteborg (June–July 1914). All parts of circulating exhibition organized by Galerie Der Sturm.

64. London. Marlborough Fine Art Ltd. *Art in Revolt: Germany 1905–25.* Oct.–Nov. 1959.

Preface and catalogue edited by Professor Will Grohmann. Section on "The spiritual in art: Der Blaue Reiter." Also German text.

65. Marbach (Neckar). Schiller-Nationalmuseum. Deutsche Literaturarchiv. *Expressionismus: Literatur und Kunst, 1910-1923.* May 8–Oct. 31, 1960.

Catalogue by Paul Raabe et al. Index, p. 340.

66. Munich. Hans Goltz Galerie. *Die zweite Ausstellung der Redaktion Der Blaue Reiter. Schwarz-Weiss.* [Feb. 12–Apr. 1912].

67. Munich. Haus der Kunst. *Der Blaue Reiter. München und die Kunst des 20. Jahrhunderts, 1908–1914.* Sept.–Oct. 1949.

Text by Ludwig Grote. Extracts from writings by Klee, Kandinsky, Marc.

68. Munich. Haus der Kunst. *München 1869–1958: Aufbruch zur modernen Kunst.* June 21–Oct. 5, 1958.

"Der Weg zum Blauen Reiter," pp. 303–44, incl. illus.

69. Munich. Moderne Galerie Heinrich Thannhauser. *Die Erste Ausstellung der Redaktion Der Blaue Reiter. München 1911–12.* Dec. 18, 1911–Jan. 1, 1912.

Exhibited Cologne (Gereonsclub).—Berlin (Sturm-Galerie).—Bremen **Bibliog-**
(Vereinigte Werkstätten für Kunst in Handwerk).—Hagen (Folkwang- **raphy**
museum).—Frankfurt (Salon Goldschmitt).

70. New York. Leonard Hutton Galleries. *"Der Blaue Reiter."* Feb. 19–
Mar. 30, 1963.
Foreword by Will Grohmann; "The universality of the Blue Rider" by
Peter Selz; "My reason for this exhibition" by Leonard Hutton Hutsch-
necker: in English, French, and German.

71. New York. Museum of Modern Art. *German Art of the Twentieth
Century.* Oct. 2–Dec. 1, 1957.
For publication, see bibl. 31, including section on "The Blaue Reiter,"
pp. 54–70, and catalogue of exhibits.

72. New York. Curt Valentin Gallery. *Der Blaue Reiter.* Dec. 7, 1954–Jan. 8,
1955.
Includes partial reprint of Kandinsky's letter to Westheim from *Das
Kunstblatt,* 1930 (see bibl. 46) and Marc's article from *Pan,* Mar. 7, 1912.

73. Paris. Musée National d'Art Moderne. *Le Fauvisme français et les
débuts de l'Expressionisme allemand.—Der französische Fauvismus und
der deutsche Frühexpressionismus.* Paris, Jan. 15–Mar. 6, 1966; Munich,
Haus der Kunst, Mar. 26–May 15, 1966.
Bilingual texts. Includes the *Blaue Reiter* within the general context, e.g.,
Kandinsky (1907–11), Macke (1910–12), Marc (1910–14).

73a. Vienna. Österreichische Galerie. *Der Blaue Reiter und sein Kreis.*
Aug. 2–Sept. 9, 1961.
Introduction by Heinrich Rumpel. 125 works also exhibited Neue
Galerie der Stadt Linz, Sept. 30–Oct. 29. Included von Werefkin and J. B.
Niestlé.

# Individual References

*Allard*
See Lankheit reference 1, p. 261.

*Burliuk*
74. *Color and Rhyme.* Editor: David Burliuk. Hampton Bays, L.I., N.Y.,
Marussia Burliuk, Publisher, 1931–65?.
Scattered references to German experiences. Also special items, e.g.,
"25 years after Mayakovsky, 1930–1950" includes section on the *Blaue
Reiter* (p. 3). See Lankheit references 2, p. 261.

*Busse*
See Lankheit references 3, p. 261.

*Hartmann*
See Lankheit references 4, p. 263.

*Jawlensky*
75. Hutton, Leonard, Galleries. *A Centennial Exhibition of Paintings by Alexej Jawlensky.* New York, Feb. 17–Mar. 1965.
Extensive documentation. Foreword by Clemens Weiler supplemented by his introduction to: *Alexej v. Jawlensky.* New York, Kleeman Galleries, Nov. 15–Dec. 1956.
76. Weiler, Clemens. *Alexej von Jawlensky, der Maler und Mensch.* Wiesbaden, Limes, 1955.
Bibliography. Supplemented by: *Alexej von Jawlensky.* Cologne, DuMont Schauberg, 1959. Includes oeuvre and bibliography.

*Kandinsky*
77. Kandinsky, Wassily. *Reminiscences.* In *Modern Artists on Art,* edited by Robert L. Herbert, pp. 20–44. Englewood Cliffs, N.J., Prentice-Hall, 1964.
New translation by Eugenia Herbert of "Rückblicke" from *Kandinsky 1901–1913* (Berlin, Der Sturm, 1913). Variant text issued 1945 by Guggenheim Museum.
78. Kandinsky, Wassily. *Concerning the Spiritual in Art.* New York, Wittenborn, Schultz, 1947.
In "Documents of Modern Art" series, edited by Robert Motherwell, including important supplementary material. Other English translations: London, Boston, 1914; New York, 1946. Central observation: "the harmony of color and form must be based solely upon the principle of the proper contact with the human soul."
79. Kandinsky, Wassily. *Essays über Kunst und Künstler.* Edited with notes by Max Bill. Stuttgart, Hatje, 1955.
Includes "Der Blaue Reiter," pp. 123–28.—"Franz Marc," pp. 185–92.
80. Grohmann, Will. *Wassily Kandinsky: Life and Work.* New York, Abrams, 1958.
Chapter III, "The Munich Years," covers 1898–1914 and the *Blaue Reiter.* Oeuvre catalogue. Extensive bibliography based on notes by B. Karpel.
81. Roethel, Hans Konrad. *Kandinsky, das graphische Werk.* Cologne, DuMont Schauberg, 1970. p. 500 (index).
Oeuvre catalogue and bibliography.
82. Solomon R. Guggenheim Museum. *Vasily Kandinsky: Painting on Glass.* (*Hinterglasmalerei*). New York, Dec. 1966–Feb. 1967.

Introduction by H. K. Roethel. Notes; 40 illus. Includes *Der Blaue Reiter* cover design in *St. George* (1911) example. See also Lankheit references 5, p. 263.

*Klee*

83. Giedion-Welcker, Carola. *Paul Klee*. Translated by Alexander Gode. New York, Viking, 1952. Bibliography.
84. Grohmann, Will. *Paul Klee*. New York, Abrams, 1954. Bibliography.
85. Haftmann, Werner. *The Mind and Work of Paul Klee*. New York, Praeger, 1954.
Original edition: *Paul Klee. Wege bildnerischen Denkens*. Munich, Prestel, 1950. Bibliography.
86. New York. Museum of Modern Art. *Paul Klee*. New York, Museum of Modern Art and Arno Press, 1970.
A reprint combining three previously published catalogues (1930, 1946, 1949), including essays and bibliography.

*Kulbin*
See Lankheit references 6, p. 264.

*Kuzmin*
See Lankheit references 7, p. 264.

*Macke*
87. Erdmann-Macke, Elisabeth. *Erinnerung an August Macke, mit einem Aufsatz von L. Erdmann*. Stuttgart, Kohlhammer, 1962.
88. Münster. Landesmuseum für Kunst und Kulturgeschichte. *August Macke: Gedenkausstellung zum 70. Geburtstag*. Münster, 1957. With five essays and catalogue.
89. Vriesen, Gustav. *August Macke*. Stuttgart, Kohlhammer, 1953, 1957.
Good documentation on artists of the *Blaue Reiter*. Bibliography. See also his preface to: *August Macke*. New York, Fine Arts Associates, Mar. 24-Apr. 17, 1952. Chronology. See Lankheit references 8, p. 264.

*Marc*
90. Lankheit, Klaus, ed. *Franz Marc im Urteil seiner Zeit*. Cologne, DuMont Schauberg, 1960.
91. Lankheit, Klaus. *Franz Marc: Katalog der Werke*. Cologne, DuMont Schauberg, 1970.
Oeuvre catalogue; exhibition list.
92. Lankheit, Klaus. Kandinsky et Franz Marc. *XXe Siècle* (Paris) no. 28, pp. 31–32. Dec. 1966.
See Lankheit references 9, p. 265.

*Rozanov*
See Lankheit references 10, p. 265.

*Sabaneiev*
See Lankheit references 11, p. 266.

*Schönberg*
93. Schönberg, Arnold. *Briefe*. Selected and edited by Erwin Stein. Mainz, 1958.
Includes Kandinsky correspondence.
94. Stuckenschmidt, H. H. Kandinsky et Schönberg. *Derrière le Miroir* (Paris) nos. 133–134, pp. 13–14. Oct.–Nov. 1962.
See also Lankheit references 12, p. 266.

# Index of Names

The index includes names from the almanac and the appendix but not from bibliographical and iconographical notes.

293